Writings of Dave Jette
2019

Pages 4-76: "A Reformulation of Dialectical Materialism" (unpublished, originally composed in 1982)

Pages 77-84: "9/11 and Nationalism" (published in the August 2018 issue of *Work in Progress*, Olympia, Washington)

Pages 85-88: "Relation of Progressives to the Democratic Party" (unpublished, November 27, 2018)

Pages 89-96: "A New Approach to Supporting Progressive Electoral Candidates" (unpublished, April 25, 2019)

Pages 97-98: Author's recent political activity

Page 99: Author's scientific career

This book is available from Lulu.com, in soft-cover ($9) and hard-cover ($19) editions.

The author, living in Seattle, can be contacted at dave@jettes.org.

Book Description

The main work in this book is "A Reformulation of Dialectical Materialism", which incorporates feminist theory into the traditional Marxist presentation of the science of dialectical materialism. Accordingly, human society is organized most broadly as an inner circle (the "Shell of Being") surrounded by a "Shell of Consciousness". The inner shell is relevant to materialism, while the outer shell concerns idealism. Each of these shells is divided into adjacent halves, with one half concerning personal interactions (individualism, informed by feminist theory) and the other with social interactions (collectivism, starting from traditional Marxist theory). Within the Personal Half-Sphere are (from center out) the categories of Biological, Personal, and Family within the Shell of Being) and those of Emotions and Rational Thought within the Shell of Consciousness. Within the Social Half-Sphere (again from center out) are the categories of Economy, Community, and Political) within the Shell of Being and those of Ideology and Science within the Shell of Consciousness. There are definite relations among all these categories which make useful an analysis of what is possible and what is not possible in influencing the development of human society.

An appendix to the main work discusses the question of whether nature exhibits dialectical development. It is demonstrated that, while life processes are indeed dialectical and indeed must be understood as such (obeying only certain of the "laws of dialectics" of traditional Marxism), nonliving phenomena (such as described by the sciences of physics and chemistry) are definitely not dialectical (in contradiction to traditional Marxist theory) and only confuse our understanding of dialectical development.

Three recent additional essays complete this book. "9/11 and Nationalism" demonstrates (based upon a review of a thoroughly convincing book) that 9/11 was an inside job rather than a foreign conspiracy, and concludes that if and when fascism comes to the United States, it will be on the basis of nationalism rather than simply white supremacy and conservative Christianity. "Relation of Progressives to the Democratic Party" and "A New Approach to Supporting Progressive Electoral Candidates" concern current electoral work.

A REFORMULATION OF DIALECTICAL MATERIALISM

Dave Jette
April 10, 2018

Dedicated to the memory of Rev. Martin Luther King, Jr.,
Fidel Castro, and Albert Einstein, each in
their own way outstanding proponents of human liberation.

Contents:

5. Preface
8 A. Introduction
12. B. The Scientific Method of Analysis
17. C. The Science of Socialism-Feminism
22. D. Are Contradictions Real?
24. E. The Elementary Dialectic
30. F. Real Contradictions
33. G. Materialist Dialectics: The Social Half-Sphere
39. H. Materialist Dialectics: The Whole Picture
    41. Figure
44. I. The Role of Feminist Theory
47. J. The Role of External Forces
50. K. Conclusion
54. Appendix: Is Nature Dialectical?
    54. A. Introduction
    58. B. Physics: Change of State of Matter
    63. C. Physics: Other Examples
    67. D. Mathematics
    71. E. Conclusion
74. References

Preface

   Traditional Marxism is wrong. In the middle of the eighteen century Marx and Engels formulated the impending socialist revolution resulting from the boundlessly increasing immiseration of the proletariat, along with the destruction of the middle classes: in order to survive, the great mass of people were going to have to rise up and overthrow the rule of the bourgeoisie. The result would be a workers' state which could eventually provide for everyone's basic needs. The working class would achieve hegemony in building the new society.

   Of course, Marxists have come a long way since then, realizing that the transformation to socialism is going to take a lot longer than anticipated because of the strength of the capitalism, and also recognizing that many struggles of a more personal nature (women's rights, LGBTQ rights, etc.) must be pursued as cross-class objects of their own, independent of the fundamental struggle between the working class and the capitalist. Yet it remains that the outlook of traditional Marxism that, in the end, it is going to be struggle between these two great classes that effects the creation of the new socialist society.

   I disagree with this outlook, and the reactionary behavior of many working-class persons (in the U.S. especially) demonstrates the weakness of this model. Rather, I follow the view of the British socialist thinker Ralph

Miliband given in this book, *Socialism for a Sceptical Age* published in 1994. In his model, as usual the state will own and control the "heights of the economy" (financial, large-scale industry, etc.), workers' collectives will own and control smaller-scale production, and there will be a major role played by small-scale private enterprise (*i.e.*, the middle classes are not going to be absorbed into the working class *en masse*.) This small-scale private enterprise will not be some sort of transitional formation, but rather a fundamental component of the new society.

So that is one fundamental break I have with traditional Marxism. The other break I have is that neither the economic class struggle nor the various personal-type struggles like for women's liberation are primary for the transformation to socialism – while integrally related to each other, they do operate independently. So, as seen in the diagram (p. 37), I have divided human activity into two parts, the Personal Half-Sphere and the Social Half-Sphere, each with its own ordering of influence of its respective components. (For example, the economy is fundamental is determining the political makeup of society, but the latter may react back on the former under certain circumstances.) It is all these interconnections and competing influences which must be taken into account in transforming to a socialist society. (Perhaps a better word for the society of the future would be "socialist-feminist".)

Furthermore, I've found it shocking that there is evidently little interest among Marxists in understanding dialectical development, for all living processes undertake such development. It's as if one wants to develop a theory of physics but eschews the understanding of advanced mathematics upon which it is necessarily based. This results partly because the outlook of traditional Marxism ("the science of Marxism-Leninism") regards the basic theory as already have been worked out by the masters, and our task is essentially to apply this theory to our particular circumstances. But another reason is that the method of dialectical materialism, as put forward especially by Engels, is dead wrong in certain of its components, producing an almost incomprehensible muddle. So the detailed appendix to my paper seeks to straighten this all out, so that dialectical materialism can be used as a tool for determining how to proceed.

A. Introduction

In order to formulate a useful theory of dialectics, we must first identify what its function is to be. In this work we use the term "socialism-feminism" for a developing science which integrates together the (complementary) Marxist and feminist approaches to understanding the collective and individual aspects of human existence, with the conscious purpose of effecting a truly liberated society satisfying the needs of its individual members. (As Marx has remarked, "the philosophers have only *interpreted* the world, in various ways; the point, however, is to change it"[1].) By a science we mean a rational, systematic exposition of the "laws" governing (or appearing to govern) a particular class of phenomena, and socialism-feminism of course differs from other sciences in the object of its study, *viz.* human society. It is the point of departure of this work that, otherwise, socialism-feminism is qualitatively the same as the other sciences, in terms of its functioning and development as a science. Accordingly, we start by defining dialectics through the function which it has roughly performed in the past: as the means of analysis, the "superstructure", for the (claimed) science of Marxism ("scientific socialism"), just as advanced mathematics is the tool which has brought about the astounding development of the science of physics. Once we have arrived at our necessary systematization of dialectics, we shall be able to apply this understanding of dialectical development to formulate a new theory of dialectical materialism which is applicable to what we are really interested in, the science of socialism-feminism.

It is the opinion of the author that dialectics has never been systematically and correctly developed as the means of analysis for the incipient science of Marxism, although very important elements of this system have been formulated by leading communist theoreticians, particularly Marx, Engels, and Mao. Especially, the practice of Marxist revolutionaries has provided the need for, and the basis of, the development of the science of Marxism, with attendant contributions toward the development of dialectics as the necessary analytic tool for the science. While Marx provided the materialist basis for dialectical analysis, it was most unfortunate that Engels attempted to systematize dialectics as governing all natural phenomena; in fact, physical (non-living) phenomena in no sense exhibit dialectical behavior, and the identification of dialectics with physics has led to a mechanical understanding of the development of human society, essentially ignoring the vital reaction of the superstructure back onto the base. Far worse, within the international communist movement under the hegemony of the Stalin group, Marxism itself was abandoned as a science, and it degenerated into a catechism used to justify (usually after the fact) the brutal economist and commandist practice of Stalinism; obviously, there was no need under such circumstances for the rectification and systematization of dialectics. A major purpose of this paper is to provide a correct formulation of dialectics, as the necessary analytic tool for the development of the science which Marxism should be helping to lead to, socialism-feminism.

But why, the reader may ask, is it important to study and systematize dialectics? Beyond its rudimentary rules, is this subject not mainly of interest to academic philosophers, and of little practical use to our work in bringing about the socialist-feminist transformation of society? This objection, well rooted in the experience of movements for progressive social change, has to be answered on a number of levels. Most fundamentally, we can never get beyond the empiricism of our movement, to develop socialism-feminism as a full-fledged science, without the "analytic superstructure" to the science which a correct formulation of dialectics can (uniquely) provide. The power of a theoretical science goes beyond the systematization of phenomena (as is carried out in its first stage of development, empiricism), to the formulation of "natural laws" which themselves necessarily conform to the "analytic superstructure" of the science. Assuming that we have a solid basic understanding of the nature of dialectical development, our attempts to understand a particular class of social phenomena, at first incomplete and containing elements of error, will be rectified and strengthened by the rigid requirements of dialectical consistency, both internal (within that class of phenomena) and external (with related classes): dialectical analysis, correctly applied, helps to sift out the wheat from the chaff in the proffered theory, and aids in seeing how to develop the theory to be more correct and more encompassing. Indeed, it can be possible in a theoretical science to develop new "laws" from (correct) old laws directly through use of the science's analytic superstructure, independent of the observation of phenomena; in this way, qualitatively new phenomena can be predicted under particular not-yet-attained conditions, as has occurred so spectacularly on occasion in the science of physics. In any case, in order to build socialism-feminism as the theoretical tool necessary for liberatory transformation, we have no choice but to develop and understand its analytic superstructure, dialectics.

Furthermore, the systematization of dialectics and the application of dialectical analysis cannot be left to the "revolutionary intellectuals" in our movement. Our vision is (hopefully) not one of the enlightened vanguard leading the relatively backward masses down the glorious path to communism (or whatever we wish to call the liberated society of the future). Rather, we must understand clearly that "communism" means the self-emancipation of the working class and people generally, and that our leadership role in this extended process must be primarily political, ideological, and theoretical (in promoting correct ways for people to act and think in transforming society) rather than organizational (in demanding that they follow plans worked out by us). This concept of leadership applies also within a socialist-feminist organization, in the relationship between its formal leadership and the rest of the organization. It is thus vitally important that all its members understand dialectics and apply it in their everyday thinking about their political work, both so that they themselves can plan and sum up their work correctly, and so that they are able to teach (by example as well as formally) people outside the organization to do the same. Although bourgeois society teaches us throughout our lives not to think dialectically, in fact social reality does develop dialectically, and in order to transform this reality we have got to understand and use dialectical analysis.

Fortunately, dialectics is not abstruse, the province of revolutionary intellectuals who spend all their lives concerning themselves with ideas. Dialectics is really quite straightforward and logical, at least in the systematization of it which we shall be presenting. We begin by examining the scientific method of analysis, in order to ascertain the role which dialectical analysis must perform in relation to socialism-feminism. Here, the relation between (theoretical) physics and advanced mathematics serves as our necessary starting point. Having ascertained what, in general, dialectics is, we must then make clear what it is not, and the appendix demonstrates that dialectics, as an abstract description of qualitative development (self-transformation), has nothing to do with nonliving phenomena, *i.e.,* with the physical sciences. Proceeding, we shall first outline a systematization of dialectics, and then we apply it for use as the "analytic superstructure" for the science of socialism-feminism.

B. The Scientific Method of Analysis

In order to understand what socialism-feminism must be as a science, and in particular the function which dialectics must play with respect to socialism-feminism, we review the scientific method of analysis as it is practiced in the science of physics, and the attendant role of advanced mathematics as a means of analysis for that science. Physics, of course, has made incredible strides in just the last century in bringing nature under our control through our knowledge of its laws. As we shall see, what it is that has enabled physics to make such astounding progress is the development and application of mathematical analysis, particularly (differential and integral) calculus but also including much more abstract mathematical techniques, such as group theory and linear algebra. It is this role played by mathematical analysis as a "superstructure" to physics which we must generalize, in identifying the corresponding function of dialectical analysis. Knowing what dialectics must do, then, we shall be in a position to give a useful formulation of this necessary means of analysis for the science of socialism-feminism.

The development of the science of physics has proceeded from practice to theory and then back again (at least ultimately) to practice. Through daily activity and conscious experimentation we have accumulated data about our environment, and we have systematized this data, step by step, into physical laws ("theory") expressed directly in mathematical terminology. The basic method for solving a particular problem in physics follows five steps:

1. The relevant data describing the particular situation under consideration is accumulated.

2. The appropriate physical laws are identified and the data is integrated into the mathematical framework of these laws.

3. Mathematical analysis is applied to the equations thereby presented – the equations are "solved".

4. The solution to the problem, expressed in mathematical language, is translated back into physical (real) language.

5. The solution is applied practically ("tested in practice"): a desired material object may be constructed, or an experiment to verify the validity of the solution may be performed.

It is only this final step, carried out immediately or at least ultimately, which establishes the validity of the physical laws (theory) being employed in the solution of the particular problem under study, and, especially, the appropriateness of the methods of mathematical analysis used in Step 3. (Otherwise, we have no reason to prefer one method of mathematical analysis to another, for all mathematical systems are logically self-consistent, and not dependent upon material reality for their existence as logical entities.)

But what are these "physical laws" used in Step 2 to handle the problem, into whose mathematical framework the data of the problem is integrated for the mathematical analysis of Step 3? "Physical laws" are what we call theory: the generalization, the abstraction, of particular observations of physical phenomena. Furthermore, physical laws are produced not only directly, through the (intellectual) process of conception; more important, for the qualitative advancement of the science of physics, is the production of new physical laws through the application of mathematical analysis to known physical laws. Such new physical laws may be relevant to situations more specific than those of the general laws started with, in which case they are arrived at through the process of deduction. Or, on the other hand, they may describe qualitatively new situations, and be arrived at through the process of conception. In the first case, the production of more specific (and detailed) physical laws tends to take the place of Step 3 above: the mathematical analysis need not be carried out explicitly for each particular problem, for theoretical physicists have usually already applied mathematical analysis to the physical laws themselves to solve in general problems which frequently arise, so that the engineer (say) can simply plug the data into the appropriate formulas (the "new physical law") to solve a problem. In the second case, the production of new physical laws applicable to qualitatively new situations enables us to identify and create new phenomena directly from the new physical laws (rather than from experimental or chance observation): for example, electromagnetic waves (radio, television, etc.), atomic energy and other phenomena predicted by Einstein's Special Theory of Relativity, lasers, airplanes.

What is happening here? Is theoretical physics, working directly with this body of physical laws produced as just described, divorced from material reality in the sense of operating independent of and separate from the data describing particular situations? The answer is yes, that in its abstraction from (material) reality lies its power, for it is only then able to use the techniques of higher mathematics developed for this purpose. The mathematical analysis used in theoretical physics is so abstract as not to be able to be related directly to the material world (in contrast, for example, to the mathematical operations of addition and subtraction, which can be viewed as the addition or subtraction of physical objects), and it furthermore is exact (at least in principle – the mathematical equations involved may be so complex as to require approximations in order to derive useful results). Therefore the physical laws of theoretical physics are considered to be exact, to be abstractions from reality to which mathematical analysis can be rigorously applied to obtained results (new physical laws or solutions to particular problems); these results in turn are exact mathematical expressions, abstractions which must be translated into material terms in order to practically useful. So in actually solving a problem in physics, we start with data which is only an approximation to reality (because of limitations on our scope and accuracy of physical measurements, and because our mathematical techniques are able to handle only a certain amount of data), apply the (exact, abstract) method of analysis of theoretical physics, and arrive at results which can be applied (again, in an approximate way) to the situation under consideration. It is this scientific method of analysis used here which we wish to generalize to socialism-feminism.

Key to understanding this scientific method of analysis is the formal exactness of the physical laws, which is possible and necessary because of the abstraction from material reality of the scientific method of analysis. Generally, physical laws have been discovered through attempts to systematize and explain observed phenomena: it has been found in practice that simple rules derived from one set of data could successfully be applied to other sets of data in the same field and be used to predict phenomena, and could even be generalized for application to related fields. Therefore, in order to make nature serve their own purposes, people have looked for and found simple physical laws which nature "obeys". In physics, the tremendously successful application of abstruse mathematical analysis has demanded that the physical laws be treated as exact abstractions to which material reality is an approximation, and when data emerges which contradicts the predictions of physical laws it is often possible to generalize the physical laws by redefining the material things to which the laws refer, so that the exactness and abstraction of the physical laws is retained. For example, when Newton's Laws of Motion were found to be in contradiction with phenomena which could be explained only by the Special Theory of Relativity, those laws were retained simply by redefining the concepts of momentum and energy in a way which reduced to the previous concepts in situations in which Newton's Laws are applicable.

We thus see that, because the physical laws are abstractions, the concepts contained within the laws, which the laws relate to one another, must also be abstractions. One can take either the (materialist) view that these concepts are approximations to material reality or the (idealist) view that material reality is an approximation to these concepts, but we note that it is the latter view which is implicitly assumed as we merrily go about applying our (abstract) analysis to the physical laws. Thus the scientific method of analysis is, formally, idealist, and in physics its validity is established beyond question by its unique ability to explain and predict material phenomena. (As noted previously, this is also the test of the validity of the particular mathematical techniques employed.) This scientific method of analysis, whose application has provided us with such great under-standing of and control over physical phenomena, must now be generalized to the revolutionary study of human society, *i.e.*, to the science of socialism-feminism.

C. The Science of Socialism-Feminism

Let us sum up what we have learned in the preceding section about the science of physics. Physics is a *science*, which means that it gives us rational knowledge of a certain class of phenomena, knowledge which both explains (systematizes) and predicts these phenomena. Using this science, we "solve" a particular physical problem confronting us through the five steps listed in the preceding section, proceeding from practice (material reality) to theory (ideal "reality") and back again to practice. Physics has developed as a science in the same way, with its "laws" being (at least originally) generalizations of observed phenomena, and requiring "testing in practice" for definitive establishment. Like any science, physics as a whole is materialist, treating material reality as *the* reality and its theory ("physical laws") as abstractions from reality which must be verified practically.

But within the science of physics, we have theoretical physics, which has as its subject matter not material reality explicitly, but the abstraction of this material reality, the "physical laws" relating abstract concepts. Along with mathematical analysis (the "superstructure" of physics), the physical laws are used to solve concrete problems, in the way which has been described. But theoretical physics also has a "life of its own", for physical laws can be derived not only as systematizations of empirical data, but also through the application of mathematical analysis to the physical laws themselves. It is the latter method of producing physical laws which accounts for the immense development of the science as a whole, for these new physical laws may be either more detailed and specific, or applicable to previously unknown phenomena – in either case, our knowledge of material reality is greatly increased through these manipulations in the theoretical domain, manipulations which are possible only because of the abstractness of the physical laws and the use of advanced mathematical analysis. While the physical laws themselves must be tested in practice, at least ultimately, theoretical physics is formally idealist, with its concepts and its laws (which are, in fact, only relationships between concepts) considered to be reality to which the external world is but an approximation.

Before generalizing the example of physics to the science of socialism-feminism, we give an indication of what is in store for us in our study of dialectics, by considering the dialectical nature of physics. (Here we shall only mention some of the concepts and relationships involved, leaving their elaboration to subsequent). The science of physics contains the following fundamental contradiction: overall, it is materialist (practice-theory-practice), but theoretical physics, which provides the motive force for its qualitative advancement, is idealist. What we say, in understanding the (dialectical) development of physics, is that this science is a dialectical unity of these two contradictory aspects: materialism is the "positive" aspect, dominant at most times and providing the quantitative development of the science, while idealism is the "negative" aspect, providing qualitative development (self-transformation) at times when it is dominant. And such it is with all sciences, *i.e.* with all the intellectual systems for the production of rational knowledge of our environment: perception (materialism) is the positive aspect, while conception (idealism) is the negative aspect. It is because of our intelligence – our ability to conceive, to abstract, to generalize – that we are able to shape and control our environment according to our needs.

In physics, the scientific method of analysis consists of the application of advanced mathematics to abstractions of physical reality: either to data expressed in the mathematical framework of (abstract) physical laws, in order to solve a particular problem (Step 3 in the five steps of the preceding section); or to physical laws themselves, in order to derive new physical laws (the method of theoretical physics). Once a science has developed its abstract concepts, its "laws" expressing relationships between the concepts, and the analytic "super-structure" necessary to manipulate the laws as well as to solve particular problems, it is able to make great progress in providing us with rational knowledge of our environment. ("Rational knowledge" is conceptual, generalized, systematized, as opposed to perceptual knowledge, which is particularized and fragmented.) In the science of socialism-feminism, dialectics provides (or will provide, once it is properly formulated) this necessary analytic superstructure.

Looking at the present state of Marxism, we see that it is still its infancy as a science. Revolutionary experience in attempting to supersede capitalism has provided us with multifold lessons for summation into theory, and individual socialist thinkers have provided us with highly important theoretical advances, but, on the whole, Marxist theory is in critical need of much development and systematization. (This is the longstanding "crisis of Marxism" which has now been clearly evident for decades.) Under the hegemony of the counterrevolution in the Soviet Union after the death of Lenin, Marxist theory has been treated as a dogma, to be used to justify whatever seems to be the thing to do at the time. The new Soviet ruling class, led by Stalin, ensured that Marxism degenerated into empiricism and economism, progressively infected with bourgeois ideology, so that there was no possibility for the development of Marxist theory. (Such development had to come from at least partially outside Stalinist hegemony, as from China under the leadership of Mao.) But by now most Marxists in the United States have become aware that there is something fundamentally wrong with what today passes for the "science of Marxism" (or of "Marxism-Leninism", the term adopted by the Stalinists).

What is needed in order for Marxism, and consequently socialism-feminism, to make major advances as a science is the application of the scientific method of analysis. Specifically, three elements are needed:

1. Abstract concepts expressing social reality.
2. Theoretical laws – the "principles of socialism-feminism" – which express relationships among the concepts.
3. A general system of abstract analysis – dialectics – appropriate to these concepts and laws.

With these three elements, we should be able to solve particular problems facing us, using steps analogous to those listed in the preceding section, and to develop socialist-feminist theory directly through application of dialectical analysis to previously known theory. This, at least, is the promise which application of the scientific method of analysis holds for the science of socialism-feminism.

We have now identified, generally, the role which dialectical analysis is to play with respect to socialism-feminism, by first understanding how socialism-feminism must function as a science. This knowledge will greatly aid us in making a useful formulation of dialectics, one fully appropriate to the requirements of this science. This we shall do in Section E. However, in order to avoid erroneous conceptions of dialectical development, we must also address in detail the question of whether nature is dialectical, and this is done in the somewhat technical appendix.

D. Are Contradictions Real?

> The fundamental cause of the development of a thing is
> not external but internal; it lies in the contradictoriness
> within the thing. There is internal contradiction in every
> single thing, hence its motion and development.
> Contradictoriness within a thing is the fundamental cause
> of its development, while its interrelationships and interac-
> tions with other things are secondary causes.[2]

Mao is here restating the classical, and fundamental, Marxist view
that all phenomena (both material and ideal) develop qualitatively
under the action of their internal contradictions, a dialectical view
which opposes the metaphysical outlook of essentially static
objects undergoing only quantitative change. In the appendix we
have shown that nonliving phenomena (the subject matter of the
sciences of physics and chemistry, as well as mathematics, which
isn't a science) do not exhibit dialectical behavior, but rather are
properly understood through cause-and-effect relationships, *i.e.* by
mechanical materialism. But, restricting ourselves to living
phenomena, and particularly to the development of human
society, is it not true that contradictions are real, in the sense of
being the *cause* of development of society? Mao states further:

> According to materialist dialectics, changes in nature are
> due chiefly to the development of the internal
> contradictions in nature. Changes in society are due
> chiefly to the development of the internal contradictions in
> society, that is, the contradiction between classes and the
> contradiction between the old and the new; it is the devel-
> opment of these contradictions that pushes society

forward and gives the impetus for the supersession of the old society by the new.[3]

In these two quotations there are two, somewhat different uses of the role of contradiction. In the first, contradiction is treated abstractly as being the cause of motion, so that, if we accept this view, then the fact of qualitative development follows naturally; qualitative development is "explained". In the second quotation, on the other hand, contradiction is being treated concretely as a tool of analysis for understanding (real) changes in society. In the previous section we made it clear that what we are after is the second use of the role of contradiction, in systematizing dialectics as the "analytic superstructure" for the science of socialism-feminism, just as advanced mathematics is for the science of physics. To put our work into perspective, however, we must still address the question of the first use of the role of contradiction: are contradictions real in the sense of being the *cause* of motion in society?

That contradiction is the cause of motion is dialectical mysticism. This view, in its unqualified form, is objectively an attack on the scientific method, for it obviates the detailed analysis of cause-and-effect relationships which is necessary for full understanding of social development. If, in one's analysis, one rejects the method of mechanical materialism in favor of complete reliance upon dialectical concepts, then one cannot grasp the particularity of the situation under consideration – one is left only with generalized (albeit valid) relationships. While we do consider dialectical analysis to be valid, and indeed absolutely necessary, in analyzing complex social phenomena, we would also say that contradiction is no more the cause of social development than the law of gravity is the "cause" of a rock's falling to the earth.

Thus we are viewing contradiction, and dialectical concepts in general, as being an idealization rather than a material cause – as a way of our understanding reality, but not as having an existence outside of our minds. This idealist conception of the role of dialectics is an integral part of the scientific method, as explained for the science of physics in Section B; if we develop dialectics as the analytic superstructure to socialism-feminism, it will have seemingly magical power to advance our science, but if we remain muddled in regarding dialectical concepts as having material reality, we shall never progress beyond empiricism. And so we proceed to develop a systematization of dialectics, and to those who object to our idealist outlook (still fighting the battle between materialism and idealism, which in science was long ago won by materialism), we can only reply that the test of our method is not your anti-scientific bias against all aspects of idealism, but rather is whether the materialist dialectics which we proffer is useful in the revolutionary work of analyzing, under-standing, and transforming social reality.

E. The Elementary Dialectic

In order to understand elementary dialectical processes, we turn once again to Mao, in discussing the question of whether the principal aspect of a contradiction sometimes loses its dominant position:

> Some people think that this is not true of certain contradictions. For instance, in the contradiction between the productive forces and the relations of production, the

productive forces are the principal aspect; in the contradiction between theory and practice, practice is the principal aspect; in the contradiction between the economic base and the superstructure, the economic base is the principal aspect; and there is no change in their respective positions. This is the mechanical materialist conception, not the dialectical materialist conception. True, the productive forces, practice and the economic base generally play the principal and decisive role; whoever denies this is not a materialist. But it must also be admitted that in certain conditions, such aspects as the relations of production, theory and the superstructure in turn manifest themselves in the principal and decisive role. When it is impossible for the productive forces to develop without a change in the relations of production, then the change in the relations of production plays the principal and decisive role. The creation and advocacy of revolutionary theory plays the principal and decisive role in those times of which Lenin said, "Without revolutionary theory there can be no revolutionary movement". When a task, no matter which, has to be performed, but there is as yet no guiding line, method, plan or policy, the principal and decisive thing is to decide on a guiding line, method, plan or policy. When the superstructure (politics, culture, etc.) obstructs the development of the economic base, political and cultural changes become principal and decisive. Are we going against materialism when we say this? No. The reason is that while we recognize that in the general development of history the material determines the mental and social being determines social consciousness, we also – and indeed must – recognize the reaction of mental on material things, of social consciousness on social being and of the superstructure

on the economic base. This does not go against materialism; on the contrary, it avoids mechanical materialism
and firmly upholds dialectical materialism.[4]

We agree entirely with Mao, but point out that he is here
challenging the mechanical materialism which enveloped the
international communist movement with the rise to state power in
the Soviet Union of the Stalin group, as justification for their
economist policies.

Attempting to take the essence of Mao's above
contribution to the theory of dialectical development, we now put
forward a model of the elementary dialectic. A situation which
experiences qualitative growth (*i.e.*, any life process, such as the
development of human society, the development of a science,
learning to walk, democratic centralism) can be considered to
have two contradictory aspects, one of which we call "positive"
and the other "negative". Most of the time the positive aspect
dominates the situation, and the situation experiences only
quantitative development, *i.e.*, its "character" ("nature") does not
change. But at times the negative aspect dominates, and the
situation undergoes qualitative transformation. The period of
dominance of the positive aspect is a stable time for the situation,
and we view the positive aspect as being the "base" and the negative aspect as being the "superstructure"; on the other hand,
during the (relatively brief) period of dominance of the negative
aspect, the situation is of course unstable and it is in the process of
becoming something else (and hence lacking clear identity). For
example, within the sphere of organization (democratic centralism), the fundamental contradiction is between democracy
(individual members' telling the group as a whole what to do) and
centralism (the group's telling individuals what to do); democracy

is the positive aspect, while centralism is the negative aspect, and it is only when the latter is dominant (when the group functions as a unified whole) that the situation can undergo qualitative change.

And that's all there is to elementary dialectical development! No mysticism about contradictions "causing" qualitative development, but the recognition that, if dialectical development does exist, we can analyze it and understand it by identifying the "positive" and "negative" aspects of the situation (or process). We have here generalized, to "positive" and "negative" respectively, the dialectically contradictory concepts of base and superstructure, or of content and form, because we need abstract concepts which are not conditioned by particular kinds of situations. Looking at a particular situation (or process) undergoing elementary dialectical development, we know that we are to look for two contradictory aspects: a "positive" one which is usually dominant, which defines the character or nature of the situation; and a "negative" one whose dominance signals qualitative transformation, changing the character of the situation.

Thus dialectical analysis provides us with a formalism for analyzing and understanding qualitative change (which does exist in society), but it is not a substitute for detailed examination of the cause-and-effect relationships operative in a particular situation. Let us conclude this section by seeing how our understanding of the elementary dialectical squares with classical Marxist thought, as summed up in Engels's well-known three basic laws of dialectics:

The law of the transformation of quantity into quality and vice versa;

The law of the interpenetration of opposites;

The law of the negation of the negation.[5]

       The first law is a bit difficult to fathom. What is meant is that quantitative changes in the situation lead to qualitative development of the situation, that qualitative development results from the continual piling up of quantitative changes. It was put forward in opposition to the metaphysical view that things don't change (qualitatively), that (bourgeois) society is eternal. Conversely, the second part of the law, transformation of quality into quantity, can be interpreted to mean that once a situation has undergone qualitative change, there arise new opportunities for quantitative change. (Such is an interpretation of this law which is favorable to Engels. In fact, he actually had in mind the transformation among themselves of certain physical quantities, such as force and energy, or mechanical motion and heat, which he identified with concepts of quantity and quality respectively. This identification is scientifically invalid, as explained in the appendix.) Engels's first law seeks to explain the existence of dialectical development as resulting from the transformation of quantity into quality (and vice versa), but it actually doesn't prove anything – it is simply the empirical observation that (at least some) situations undergo qualitative development, first only changing quantitatively but then actually transforming themselves qualitatively. Similarly, our elementary dialectic is a description of the observed fact that situations undergo qualitative change, but we make no pretense of understanding the cause of qualitative change, at this level of analysis.

       Engels's second law, the mutual interpenetration of opposites, was for Mao the fundamental law of dialectics, that contradiction lies in the very essence of things (and hence that all things undergo dialectical development). As interpreted by Mao, this law describes the motive force of qualitative development, that internal contradiction *causes* such development, which we already have rejected, but the law is of fundamental importance in analyzing real contradictions, as discussed in the next section. In terms of our elementary dialectic, we could say that the positive and negative aspects interpenetrate each other in the sense that one cannot exist without the other (just as content and form cannot exist without each other), and that there exists "struggle" between the two for the position of dominance, but it is difficult to see the benefit in making such a statement.

Finally, Engels's third law, the negation of the negation, fits right into our elementary dialectic: the negation of a situation occurs when its "negative" aspect gains dominance, qualitatively transforming the situation (relatively quickly) into the "negation of the negation", in which the "positive" aspect regains dominance and the situation stabilizes, to undergo quantitative change until its next negation. (Indeed, Engels's third law is what has led us to the selection of the terminology "positive" and "negative".) We do have to be very careful in using Engels's third law, however, for it has been so badly misused by Marxists as to become a meaningless mechanical exercise.

For example, we view the fundamental contradiction of class societies as being that between the economic base and the superstructure (consisting of "everything else" social: the half-shells of community, politics, ideology, and science, as discussed in Section G). It is because it is our primary concern as living organisms to satisfy our material needs, that the economic base constitutes the "positive" aspect and the superstructure the "negative" one. At most times the economic base is dominant in "determining" the various elements of the superstructure ("historical materialism"), and we have a (relatively) stable mode of production. But there will arise times when the superstructure attains dominance and transforms society, thereby leading to a new mode of production. The period of socialism is a time when such a qualitative change is taking place, when the superstructure is dominating the development of society in its transformation to the communist mode of production. (More on this in Section J.) But we would not say that capitalism is the negation of feudalism (as is often suggested), for both are modes of production – rather, capitalism must be viewed with respect to feudalism as a "negation of the negation", just as communism is a "negation of the negation" of capitalism.

F. Real Contradictions

Mao, in "On Contradiction", has written at great length about the analysis of social contradictions, in order to understand how society changes and how best to intervene in its development. We consider this work to be of fundamental importance. However, it does mix up elementary dialectical processes with the real (complex) processes which actually occur in social development, which involve contradictions between real objects and which are manifestations of whole complexes of interconnected dialectical processes. Mao is most concerned with this second type of process, as is seen in the following:

> In speaking of the identity of opposites in given conditions, what we are referring to is real and concrete opposites and the real and concrete transformations of opposites into one another.[6]

We term contradictions constituting an integral part of such processes, "real contradictions".

In order to understand the difference between real contradictions and the contradictions which occur in the (ideal) system of materialist dialectics which we shall be developing in the next section, we consider the following passage from "On Contradiction":

> There are two states of motion in all things, that of relative rest and that of conspicuous change. Both are caused by the struggle between the two contradictory elements contained in a thing. When the thing is in the first state of motion, it is undergoing only quantitative and qualitative change and consequently presents the outward appearance of being at rest. When the thing is in the second state of motion, the quantitative change of the first state has already reached a culminating point and gives rise to the dissolution of the thing as an entity and thereupon a qualitative change ensues, hence the appearance of a conspicuous change.[7]

Mao is here describing the sequence of quantitative change leading to qualitative change, but not in terms of the "positive" and "negative" aspects of the dialectical process, as we did in the previous section. Rather, he is speaking of the mutual struggle of two real contending elements of the thing, without placing them in the position of positive and negative aspects. Mao does concern himself with which aspect of a contradiction holds dominance (the "principal aspect"), but his dialectic is different from our elementary dialectic:

When the principal aspect which has gained predominance changes, the nature of a thing changes accordingly.

In capitalist society, capitalism has changed its position from being a subordinate force in the old feudal era to being the dominant force, and the nature of society has accordingly changed from feudal to capitalist. In the new, capitalist era, the feudal forces changed from their former dominant position to a subordinate one, gradually dying out.[8]

In contrast, we would not identify the forces of feudalism and the forces of capitalism with the two aspects of an elementary dialectic. To mix up real contradictions with the elementary dialectic in this way would render the law of the negation of the negation meaningless, and Mao is at least true to his theory, for he realizes that in it the law of the negation of the negation is not valid.[9]

In making clear the distinction between our materialist dialectics and the real contradictions with which Mao is chiefly concerned in "On Contradiction", it is not our purpose to denigrate Mao's substantial contribution to Marxist philosophy. To the contrary, Mao's work on identifying the principal contradiction and the principal aspect of a contradiction, as well as on understanding the role of antagonism in contradiction, is of the highest importance in analyzing real dialectical movement. Our point is that our work is somewhat different, that in propounding our materialist dialectics (in the next section) we are attempting to provide a theoretical framework within which the real contradictions of society can be analyzed, understood, and responded to.

G. Materialist Dialectics: The Social Half-Sphere

We now have our elementary dialectic, the basic building block for constructing our analytic superstructure to the science of society. We have already demonstrated, in Appendix A, that nonliving phenomena do not exhibit dialectical behavior, but have left open the question of whether all life processes are dialectical. In our opinion the answer is yes, and one could construct a dialectical superstructure to the science of biology, for example. However, we would against so doing, because it would not really be useful in advancing the science of biology. The subject matter of biology is simple enough to be well understood by mechanical cause-and-effect relationships (mechanical materialism), and even elementary life processes can be analyzed and understood through the (sub-) science of biochemistry. On the other hand, the development of human society is extremely complex, so complex that a dialectical superstructure is useful, indeed necessary, to any science of society, in order to properly relate among themselves the myriad interconnected processes occurring in a given social formation. We call this dialectical superstructure "materialist dialectics", but before proceeding with its construction we reiterate its two basic limitations. First dialectical analysis cannot take the place of the analysis of cause-and-effect relationships; both kinds of analysis are necessary for the full understanding of a particular situation or process, and relying entirely upon the former leads one straight into dogmatism. And second, dialectical analysis is only that – a tool for analyzing and understanding reality, which, like any other analytic superstructure to a science, exists only in the mind; to try to "explain" or "prove" the existence of qualitative development through such statements as that contradiction is the (material) cause of motion and development, is to indulge in mysticism.

In dealing with a complex situation or process, involving a number of interconnected subprocesses, we understand a particular dialectical process in terms of the concepts of core, base, superstructure, and atmosphere. Here, "base" and "superstructure" refer to the "positive" and "negative" aspects respectively of the elementary dialectic (contradiction) under consideration – we find the full four-part terminology necessary for analysis of complex situations. "Core" refers to the set of processes which lie "dialectically below" the contradiction under consideration, determining the milieu in which the latter develops (*i.e.*, the bounds beyond which the latter cannot develop). "Atmosphere" refers to the set of processes which lie "dialectically above" the given contradiction, which in themselves have little influence upon the core, base, and superstructure, but rather are "determined" by the latter (although contradictions in the atmosphere do have a life of their own, within the limits set by the structures "dialectically below"). Thus we have, in relating among themselves the multiplicity of contradictions of society, the concept of dialectical level of influence, with lower-level contradictions "determining" (in a dialectical sense, *i.e.*, setting the bounds for development of a contradiction) higher-level contradictions.

Before getting completely lost in abstraction, however, let us come back to reality by constructing materialist dialectics. In this section we shall make this construction for the "social half-sphere", for which Marxist theory is useful, and in the next section we shall present the full picture by bringing in the "personal half-sphere", for which feminist theory comes into play. The figure on page 41 depicts all these relationships of our new theory of dialectical materialism.

We are materialists. We understand that being determines consciousness, in the dialectical sense explained by Mao. (See the first quotation of Section E.) Thus we divide human existence (or activity) into two great shells, being and con-sciousness, which are dialectically related: being is at the lower dialectical level, consciousness at the upper level. (As seen in figure, the "shell of being" is actually a full sphere. Also, we are now looking only at the "social half-sphere", so these are actually "half-shells".) Furthermore, within the great-shell of being what is primary is satisfying our material needs, and we place the half-shell of economics (the "economic base") at the very lowest dialectical level for the social half-sphere, "determining" not only what happens in the rest of the social half of the great-shell of being, but also, through its determination of the great-shell of being, what happens in the social half of the great-shell of consciousness. The remainder of the social half of the great-shell of being we divide into the community and political half-shells, with the former lying dialectically below the latter. (The political half-shell occupies the highest level in the social half of the great-shell of being because it concerns ways of doing things, and thus corresponds to form, whereas the economic and community half-shells correspond to content. The community half-shell includes our various social interactions with one another, at the level of the community.) Finally, we divide the social half of the great-shell of consciousness into two half-shells, the ideological half-shell and the scientific half-shell, with the former at the lower level. The ideological half-shell is actually the reflection in our consciousness of our social being, and it includes our emotions, our irrational ("un-thought-through") responses to social situations we face. The scientific half-shell, in contrast, includes our rational processes (including production and use of scientific theory), and its location at the highest dialectical level of human social existence is a statement of the fact that Human, despite delusions to the contrary, is basically an irrational being.

Such is our formalism for understanding social existence (or activity): we divide it into five half-shells, *viz.* economic, community, political, ideological, and scientific half-shells, in that (dialectical) order. This division is somewhat arbitrary, and in some applications of dialectical analysis we shall want to combine half-shells; for example, we may be looking at the dialectical interaction of the economic and community half-shells, taken as a whole (the "base") with the political half-shell (the "superstructure"). More often, we may want to divide a particular half-shell into sub-half-shells for dialectical analysis; for example, we divide the economic base into the forces of production (lower level) and the relations of production (upper level). While the division is thus somewhat arbitrary, chosen to suit the particular dialectical processes being analyzed, the order of division is not arbitrary. The order of the half-shells or sub-half-shells is a statement of dialectical determination, and this ordering is determined by observation of social reality, which does indeed behave dialectically.

For example, the economic half-shell (the "economic base") comprises part of the core of all the other dialectical processes, and thus determines (at least in part) the bounds beyond which these other processes may not step without altering the very character of the economic sphere. Conversely, a drastic upheaval in the economic half-shell (a change in its nature, *i.e.* a change in the mode of production) implies a drastic upheaval in the rest of social existence, by changing the bounds of those other contradictions. Thus one cannot have a "peaceful" transition from capitalism to communism (or from feudalism to capitalism), involving only political reform; what must occur is a drastic upheaval throughout the superstructure of society. Another way of stating this conclusion is that, relative to the contradiction between the forces and relations of production ("base" and "superstructure" here), dialectical processes restricted to the political half-shell lie in the atmosphere, and they may proceed (within bounds determined by the economic and community half-shells) merrily on their own, without fundamentally affecting the economic half-shell. Political reformism is not a viable strategy for social transformation!

Althusser[10] has made a major contribution to Marxist philosophy by attacking the Hegelian view that human history is the manifestation (realization) at continually higher levels of a single essence: human mastery over the environment (hence, the development of the forces of production). The Hegelian dialectic leads to economic determinism, regarding all contradictions in society as being subordinate to the fundamental contradiction in the economic half-shell (between the forces and relations of production), as not having an independent existence of their own. Althusser's solution has been to counterpose the "overdetermination of contradiction", by which is meant the existence of a multiplicity of contradictions (dialectical processes), no one of which is the essence of which all the rest are but manifestations. The problem with Althusser's work (aside from its being well-nigh incomprehensible) is that, while it correctly poses the existence of independently developing contradictions, it does not fit them into a necessary structure relating them with each other. The concept of dialectical determination is lost, and with it, historical materialism.

It is the purpose of our proffered materialist dialectics to provide an analytic tool which both avoids economic determinism (so that the relatively independent nature of dialectical processes is recognized) and orders the processes in a determinant way (so that their necessary interconnectedness with one another is made clear). For example, one cannot have a dialectical process which involves only the economic half-shell (as base) and the political half-shell (as superstructure), for the community half-shell is trapped between the two; indeed, a political revolution which transforms the economic base must involve struggles in the community (against the various kinds of oppression of capitalist society, such as racial discrimination). Similarly, the political half-shell can change its character without qualitatively affecting the two half-shells "below" it (economic and community), as happened in Japan after World War II, and the community and political half-shells can change character together while leaving the economic half-shell basically unchanged (*e.g.*, the transformation of colonies into neo-colonies), but a revolutionary transformation of the economic half-shell, in order to be lasting, must be accompanied by revolutionary transformation of the other half-shells.

H. Materialist Dialectics: The Whole Picture

A fairly recently published book, **Liberating Theory**[11], examines in great detail the three half-shells of the social half of our "being" great-shell, *viz.* the economic, community, and political half-shells. Directly attacking the economic determinism of Classical Marxism (as developed by the international communist movement under the hegemony of Stalinism), the book demonstrates the relative independence of these half-shells and examines their internal struggles, as well as presenting the myriad connections among the half-shells. Particularly important is the identification, in the economic half-shell, of the "coordinator class" as the third major class in industrialized society, along with the capitalist and working classes; this class is discussed in some detail in its Chapter 3. Unfortunately, in its eagerness to attack the economism of Classical Marxism (including that of social democracy), **Liberating Theory** does away with the ordered dialectical relationship among these half-shells which we have just outlined, and therefore, while being very useful in understanding exactly what these half-shells are on an empirical level, it forgoes the possibility of developing a scientific understanding of social development, at a higher theoretical level.

Liberating Theory does, however, go beyond the structure of materialist dialectics which we have thus far presented here, in identifying a fourth major area of human existence, termed "kinship". (The book ignores our ideological and scientific half-shells.) The kinship area includes close interpersonal relationships based on family ties and on sexual activity, and in our society (in fact, throughout the world) its salient characteristic is male supremacy. We agree with **Liberating Theory**'s contention that the kinship area lies outside of any fixed base/superstructure relation involving the other half-shells – patriarchal society transcends the succession of economic modes of production. Evidently the kinship area belongs to the realm of "personal existence", in contrast to the realm of "social existence" to which we have assigned the other five half-shells. This insight leads to the creation, in our new theory of dialectical materialism, to a "personal half-sphere" which fully complements the "social half-sphere" which we have already presented. The integral, dialectical relationship between the two half-spheres implies a corresponding relationship between (corrected) Marxist theory and feminist theory. Indeed, this relationship implies a scientifically correct characterization of activists seeking human liberation: *socialist-feminists* (and hence the title of this paper!).

As seen in the main figure, we envision the sphere of human existence as being comprised of two half-spheres, "personal" and "social". To each of the five half-shells which we have identified as together constituting the social half-sphere, there is a corresponding half-shell – at the same level – in the personal half-sphere. These new half-shells have exactly the same dialectical relationships among themselves, as we have already explained for the half-shells in the social half-sphere. Thus the personal half-sphere has three half-shells in its "being" portion ("great half-shell") and two half-shells in its "consciousness" portion. The core half-shell of the personal half-sphere concerns biological reproduction, the next half-shell is that of personal interactions, and the "being" portion of this half-sphere is completed with a half-shell concerned with the organization of

Personal Half-Sphere   **Social Half-Sphere**

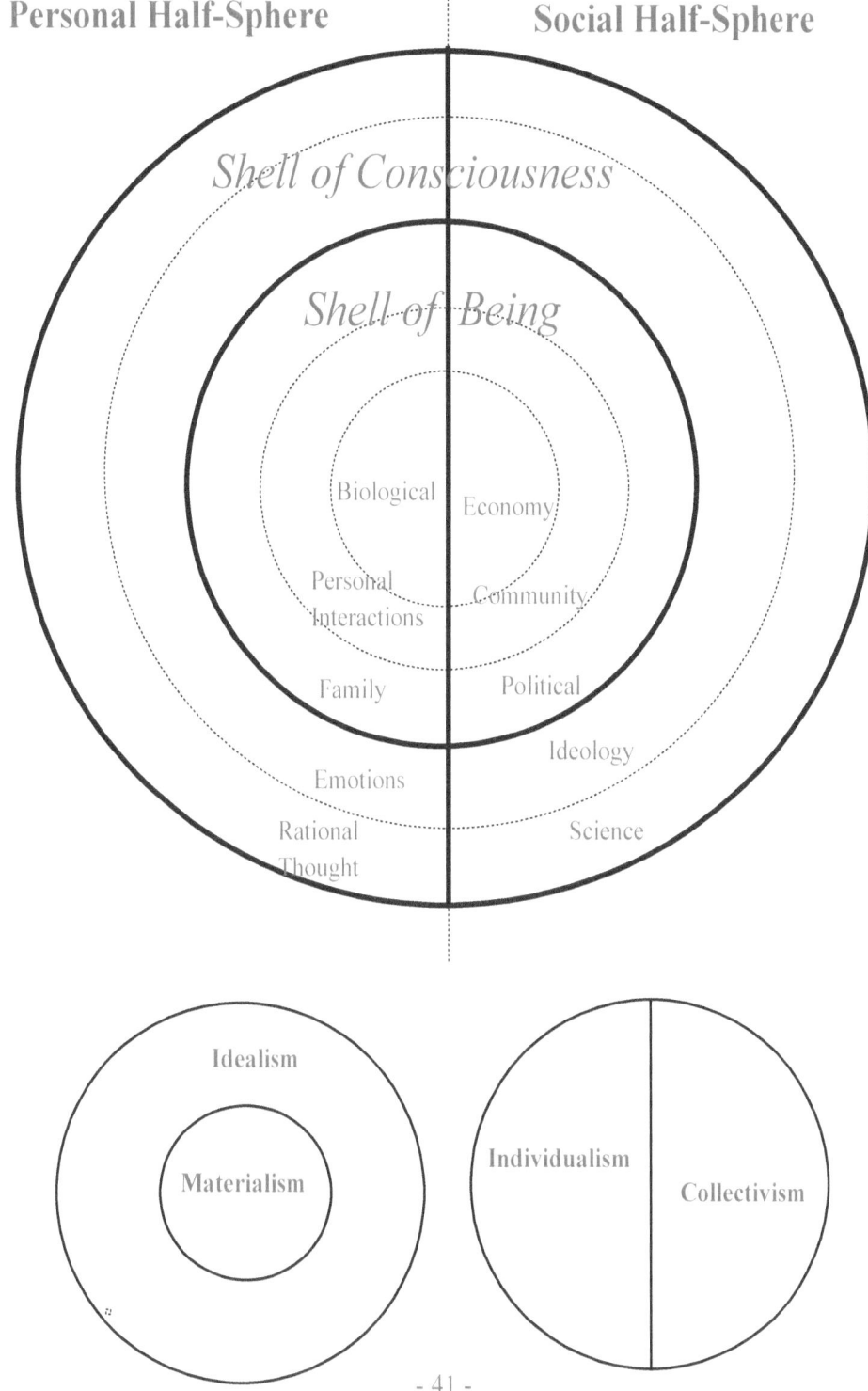

Shell of Consciousness

Shell of Being

Biological   Economy

Personal   Community
Interactions

Family   Political

Ideology

Emotions

Rational   Science
Thought

Idealism

Materialism

Individualism
Collectivism

personal life (particularly the structure of the family and other institutions of male supremacy at the personal level). The "consciousness" portion of the personal half-sphere is comprised of half-shells concerning emotions (values, religious views) and rational thought, respectively. (We should note that, within the personal half-sphere, the actual names of the half-shells are somewhat imprecise, for these half-shells can actually pertain either to individuals or to family-type units. But the content of these half-shells should be clear, if one keeps in mind the corresponding half-shells in the social half-sphere.)

Thus this structure for our theory of dialectical materialism closely integrates the personal and social aspects of human existence at all levels. The "sphere of human existence" can be considered to consist of five (full) shells, each of which consists of one half-shell each in the personal and social half-spheres (which themselves constitute a dialectical whole, of human existence). Moreover, there is a definite dialectical relationship between the two half-shells in a (full) shell, the same one as between the personal and social half-spheres: in terms of the systematization of dialectics given in Section E, the personal half-sphere is the "positive" aspect of this dialectical unity, and the social half-sphere is its "negative" aspect. (When people act together collectively, the situation can undergo qualitative change – this is the basic reason for this assignment of positive and negative aspects of this dialectical unity.)

At the emotional/ideological level, the dialectical unity between the personal half-sphere and the social half-sphere is precisely that between individualism and collectivism, and we have indicated this relationship in one of the sub-figures. The critical point here is that *individualism and collectivism form a dialectical unity with individualism being the positive aspect.* This understanding is in opposition to Classical Marxism and other collectivist ideologies, which cringe at the thought of individual demands upon the collective body; lacking scientific understanding of human existence at this theoretical level, Classical Marxism cannot lead to human liberation, and naturally falls into the trap of dictatorship and suppression of individual rights for the cause of The Revolution. At the organizational level within any of the three shells of the great-shell of being, this dialectical unity is that between democracy and centralism (with democracy constituting the positive aspect). The fundamental importance of democracy in building a socialist-feminist society is discussed in the next section.

There is another major dialectical unity evident in the main figure: that between the great-shell of being and the great-shell of consciousness. This implies the dialectical unity of materialism and idealism, and again we have a critical point in opposition to Classical Marxism: *materialism and idealism form a dialectical unity with materialism being the positive aspect,* as depicted in the other sub-figure. By failing to grasp the necessary role of idealism (as seen, for example, in the way theoretical physics functions, as explained in the appendix), Classical Marxism easily lapses into (undialectical) mechanical materialism and, in particular, economism. (We are here using "materialism" and "idealism" not in the sense of values, but rather as used in philosophy: as indicated at the end of the first section of the appendix, materialism is the view that ideas are approximations to material reality, and idealism is the view that material reality is an approximation to ideas.) One grievous implication of Classical

Marxism's blanket rejection of idealism is its rejection of human spirituality (another reason that Classical Marxism gets nowhere in its quest for human liberation.

I. The Role of Feminist Theory

We thus have a structure for a new theory of dialectical materialism. The various components of the sphere of human existence have well-defined dialectical relationships among themselves, as already indicated. Using the elementary dialectic ("positive" and "negative" aspects of a dialectical unity), as well as the concepts of core, base, superstructure, and atmosphere, we have drawn the limits of the relative independence of these components and of their ability to affect each other. This is quite different from the outlook of **Liberating Theory**, that the basic areas influence each other without any particular order of precedence. Our claim is this: if you are trying to analyze a particular situation or process and cannot fit your understanding of it into this structure of dialectical materialism, with the attendant restrictions on what can influence what and to what degree, then you are doing something wrong and should start over in your analysis. Thus we are saying that this theory of dialectical materialism is not only *useful* in describing human existence, by providing a neat way of categorizing the various aspects of a situation or process; it is *necessary* in limiting suggested interpretations of reality to those which can possibly be valid. This is what we must demand of an "analytic superstructure" to a science.

One implication of this theory of dialectical materialism is that people seeking human liberation must work to transform both the personal and social half-spheres simultaneously. It is just as meaningful to say that the struggle against capitalism will continue under a feminist organization of society, as it is to say that gender equality and women's liberation will continue as an independent struggle under socialism, for in fact the two struggles are indivisibly intertwined. (Corrected) Marxist theory is of fundamental importance to the social half-sphere, and feminist theory is similarly important to the personal half-sphere, but a complete understanding of the human condition has to be rooted in both theories.

We have in fact presented only the skeleton of dialectical materialism, and that starting from a Marxist perspective and then generalizing to include bare rudiments of feminism. To be routinely useful, this new theory of dialectical materialism will have to be fleshed out through application to their political work by social activists. Especially, it will be important that feminist theoreticians explore how existing feminist theory can fit into this structure, for it doubtlessly can be substantially refined when addressed from that perspective. Indeed, we realize that there is a whole body of feminist theory developed from the late 1960's on which must be so addressed, but which is beyond the scope of this paper (and of the author's present expertise). However, two quota-tions, from Margaret Randall and Rosalind Petchesky, indicate the compatibility between the ideas presented in this paper and feminist theory:

> I believe that in each of the revolutionary experiments the failure to develop an indigenous feminist discourse and a vital feminist agenda impeded the consolidation that would push an otherwise more humane society forward. Again and always, the men had their way. It

was obviously too frightening to have to confront the unleashed power of women, silenced for centuries. Easier to pay lip service to "women's rights", while being careful not to allow them to become a priority or to threaten male power. An important degree of stagnation set in perhaps – among other reasons – because *our very understanding of the relationship between base and superstructure was erroneous.* Perhaps the concept of *base* itself must be redefined.[12] [emphasis in the original]

[Gayle] Rubin, in her proposal[13] for a "political economy of sex-gender", thoughtfully suggested to us that sexuality and gender, and the kinship-family structures in which they are reproduced, "are themselves social products", that they consist of historically determined relationships in which material production, wealth, exchange, power, and domination – as well as feelings and sensibilities – are all directly involved. This, in turn, led to a further analytic insight: that "production" and "reproduction", work and the family, far from being separate territories like the moon and the sun or the kitchen and the shop, are really intimately related modes that reverberate upon one another and frequently occur in the same social, physical, and even psychic spaces. This point bears emphasizing, since many of us are still stuck in the model of "separate spheres" (dividing off "woman's place", "reproduction", "private life", the home, etc. from the world of men, production, "public life", the office, etc.). We are now learning that this model of separate spheres distorts reality, that it is every bit as much an ideological construct as are the notions of "male" and "female" themselves. Not only do reproduction and kinship, or the family, have their own, historically determined, products, material techniques, modes of organization, and power relationships, but *reproduction and kinship are themselves integrally related to the social relations of production and the state; they reshape those relations all the time.* One implication of this theoretical breakthrough (and I don't think that's too grandiose a term) is that the two tasks of analyzing patriarchy and analyzing the political economy – whether capitalist, pre-capitalist, or socialist – cannot be separated. The very process of developing a Marxist-feminist mode of analysis will necessarily deepen the Marxist dialectic and enrich its ways of seeing and reflecting the world.[14] [emphasis added]

J. The Role of External Forces

Mao has made clear the role of external forces, in the context of his conception of internal contradictions as the driving force of development:

Does materialist dialectics exclude external causes? Not at all. It holds that external causes are the conditions of change and internal causes are the basis of change, and that external causes become operative through internal causes.[15]

We close this discussion of dialectical materialism by examining the role of external forces, with particular application to an understanding of the proper role of leadership.

It is of the utmost importance to understand that situations develop qualitatively only when their negative aspect becomes dominant, and to understand therefore how one can best affect the course of development in a situation, as an external force. There are two modes of development of a situation, which we may term "natural" and "conscious". In "natural development", the negative aspect hinders the quantitative development of the situation provided by the dominance of the positive aspect, until finally the situation seems to explode and to undergo qualitative change under the dominance of the negative aspect, with transformed content of the negative aspect. The classical example of natural development is the cataclysmic upheaval which occurs when the (gradual) development of the productive forces "bursts the fetters" of the (outmoded) relations of production (and of the superstructure of society in general).

"Conscious development", on the other hand, results from the action of an external force upon the negative aspect of a situation, to transform the content of the negative aspect so that the negative aspect will dominate and transform the situation in the desired way. The situation still undergoes self-transformation resulting from its internal contradictions (one of the basic laws of dialectical materialism), but this transformation is consciously brought about. Thus it is the proper role of leadership (which is both integrated into and independent of the group as a whole), of external intervention, to act upon the negative aspect of a situation (its *relationships*, its *forms* of activity, rather than the activity itself) in order to bring about the self-transformation of the situation. Correct leadership is *political* leadership, putting forward ideas on ways to do things in order to release the initiative of everyone in the group, rather than trying to organize people's work for them.

Economists (by which term we mean people who systematically make errors of economism, *not* people who are involved in the study of the economy) and socialist-feminists have a fundamentally different understanding of (dialectical) development and of the proper role of external forces. ("Socialist-feminist" here refers specifically to an activist who is following the ideas presented in this paper. This is a gross idealization, of course, and in using the term in this way we are not trying to define who the "real" socialist-feminists are.) Seeing the dominance of the positive aspect of a (stable) situation, economists consider the focus of their activity to be influencing that aspect, and thereby restrict themselves to the "natural" development of the situation. Thus comes their overemphasis on trade union work and on mass democracy (before society collapses of its own weight), and their complete concentration on the development of the productive forces (after attaining political power). In contradistinction, socialist-feminists who are following the ideas presented in this paper emphasize the conscious element, the role of leadership, the need for a centralist political party, the critical role of class struggle; in general, such socialist-feminists comprehend the need to transform the content of the negative aspect (the superstructure of society, relations of production, etc.) in order to bring about the qualitative transformation of a situation. Thus, these socialist-feminists understand the possibility of the "conscious" development of a situation, and accordingly direct their forces to influence the negative aspect of the situation.

Socialism is viewed as a mode of production by economists, and as a transitional society by socialist-feminists following the ideas presented in this paper. Viewing socialism as a mode of production means considering such a society to be (relatively) stable, with the economic base constituting its positive aspect, dominating the development of society under the action of the particular "economic laws of socialism"; thus one's main concern must be the development of the productive forces. On the other hand, viewing socialism as a (relatively unstable) transitional society between capitalism and communism (or whatever one calls the liberated society of the future) entails continual concentration on the negative aspect (political leadership, class struggle, the relations of production, other social relationships, the superstructure of society in general) in order to transform the content of the negative aspect and thereby to bring about society's self-transformation to communism.

K. Conclusion

In order to work effectively toward a liberated society, socialist-feminists must break clearly with Classical Marxism, while seeking to retain its valid insights. We hope that in this paper we have clarified the weaknesses and strengths of Classical Marxism, and identified the fundamental problems with this ideological system. Objectively, and often subjectively, Classical Marxism serves the interests of a new class in modern society, the coordinator class, as explained in **Liberating Theory**; without this key realization, it may be difficult to understand why its adherents can be both genuinely opposed to capitalism and acting against the struggle for human liberation. It is important for socialist-feminists to work with adherents of Classical Marxism in particular struggles, just as we do with liberals, but we must not let them or their ideology dominate the movement. Especially, it is important to reject the top-down leadership methods of Classical Marxism, in both its "Stalinist" and social-democratic forms. We must build organizational structures and institutions which are directly controlled by, and serve the interests of, the great majority of people. The "communist" society of the future is one in which such democratic control is universal, and we must always keep this goal in mind as we strive to transcend present society.

The objective of this paper has been the development and presentation of a coherent, valid system of dialectical analysis, and particularly an initial formulation of its application to the development of society, "dialectical materialism". It is most unfortunate that, at least in the United States, social activists have often been content to go about their (political) business without concerning themselves with the structure of their thinking processes for formulating their strategy and tactics. Such pragmatism certainly allows them to proceed with the task at hand, but lack of a scientific analysis means that they often end up spinning their wheels, or worse. Without a solid understanding of dialectical development, it is impossible to grasp such fundamental dialectical unities as materialism and idealism, individualism and collectivism, and democracy and centralism. And without understanding such dialectical unities, one cannot effectively engage in working out a strategy for human liberation.

Writing overt a half-century ago, Albert Einstein emphasized the need for the replacement of capitalism with socialism, while assuring true democracy with respect for individual rights:

I am convinced there is only *one* way to eliminate these grave evils [of capitalist society], namely through the establishment of a socialist economy, accompanied by an educational system which would be oriented toward social goals. In such an economy, the means of production are owned by society itself and are utilized in a planned fashion. A planned economy, which adjusts production to the needs of the community, would distribute the work to be done among all those able to work and would guarantee a livelihood to every man, woman, and child. The education of the individual, in addition to promoting his own innate abilities, would attempt to develop in him a sense of responsibility for his fellow men in place of the glorification of power and success in our present society.

Nevertheless, it is necessary to remember that a planned economy is not yet socialism. A planned economy as such may be accompanied by the complete enslavement of the individual. The achievement of socialism requires the solution of some extremely difficult socio-political problems: how is it possible, in view of the far-reaching centralization of political and economic power, to prevent bureaucracy from becoming all-powerful and overweening? How can the rights of the individual be protected and therewith a democratic counterweight to the power of the bureaucracy be assured?[28]

This is exactly what we see as being the goal of socialist-feminists, and, most unfortunately, certain major tenets of Classical Marxism stand in opposition to this goal and must be consciously rejected. Furthermore, in contrast to the economic determinist outlook of Classical Marxism, we do not struggle to overthrow capitalism because it is preordained to collapse of its own weight, through the increasingly intense contradiction between the forces and relations of production. Rather, we seek this transformation of society because capitalism is *evil*: in its mature stage of development, it is fundamentally opposed to all that is good in humanity, both generally and in ourselves as individual beings. We of course do not forswear a scientific analysis of the internal contradictions of capitalism and its intrinsic limits, but we must also bring to the fore our own spirituality.

APPENDIX: Is Nature Dialectical?

A. Introduction

Is all nature dialectical? That it is, was a fundamental premise of Engels in formulating dialectical materialism. In the preface to the second edition of **Anti-Dühring**, Engels writes that he had spent the best part of eight years in gaining knowledge of mathematics and natural science, for such knowledge "is essential to a conception of nature which is dialectical and at the same time materialist"[16]. Thus Engels sought to understand the dialectical development of human society in terms of universal laws applicable to all aspects of nature:

> ... my recapitulation of mathematics and the natural sciences was undertaken in order to convince myself also in detail – of what in general I was not in doubt – that in nature, amid the welter of innumerable changes, the same dialectical laws of motion force their way through as those which in history govern the apparent fortuitousness of events; the same laws as those which similarly form the thread running through the history of the development of human thought and gradually rise to consciousness in the mind of man; the laws which Hegel first developed in all-embracing but mystic form, and which we made it one of our aims to strip of this mystic form and to bring clearly before the mind in their complete simplicity and universality.[17]

Engels wrote at considerable length about dialectical development in two major works: **Anti-Dühring** ("AD") and **Dialectics of Nature**[18] ("DN"). In DN (p. 26) he formulated the three basic laws of dialectics, abstracting them from "the history of nature and human society":

> The law of the transformation of quantity into quality and vice versa;
> The law of the interpenetration of opposites;
> The law of the negation of the negation.

Engels considered these to be "the most general laws of these two aspects [nature and human society] of historical development, as well as of thought itself", and he gave many examples and analyses of the manifestation of dialectical development in the physical sciences (physics, chemistry) and in mathematics, in AD and especially in DN.

We address in this paper the question of whether dialectics is applicable to the physical sciences and mathematics. Although a fundamental premise for Engels in formulating dialectical materialism, the applicability of dialectics to nature and mathematics has been essentially ignored by other classical elaborators of Marxism such as Marx, Lenin, Stalin, and Mao, who of course had other questions to occupy their time. Marx deferred to Engels's expertise in such matters, and Stalin quoted Engels to establish the universality of dialectical development[19]. Lenin made brief mention of dialectical contradiction in nature and mathematics:

> In mathematics: + and -. Differential and integral.
> In mechanics: action and reaction.
> In physics: positive and negative electricity.
> In chemistry: the combination and dissociation of atoms.
> In social science: the class struggle.[20]

And Mao quoted this passage to illustrate the universality of contradiction[21]. There have been fairly recent attempts to establish the universality of dialectics, notably by the (now-defunct) Communist Labor Party, whose journal we shall quote later in connection with the application of dialectics to mathematics. But in addressing the question of whether dialectics is applicable to the physical sciences and mathematics, we must concentrate upon the two fundamental works of Engels, AD and DN.

The natural sciences have always been "bourgeois", in the sense that they completely ignore Engels's view that their subject matter exhibits dialectical development. There was, however, a major attempt made in the Soviet Union to reformulate the natural sciences in terms of dialectics, as pointed out by Holubnychy, summarizing the study of Joravsky[22]:

> Engels left materialistic dialectics as a more or less developed philosophy, but not a science or a formal scientific method. Yet, in the twenties, Russian Communist "believers" tried to force dialectical method on all sciences, including the natural sciences. The attempt failed dismally, and dialectics became seriously compromised as a result. Tacitly, the Russians concluded that it was of little practical value in education and that it should be replaced by the study of traditional logic.[23]

There is a very basic reason that this attempt to apply dialectics to the natural sciences fell flat on its face, so that Soviet scientists had to revert back to "bourgeois science": nonliving nature is not dialectical. To establish this fact, we shall examine in detail Engels's claim that the science of physics, as well as mathematics, exhibits dialectical behavior.

In examining Engels's work, however, we are faced with a problem: if we deny his understanding of dialectics, then what can we say that dialectics is? If we don't know what dialectics is, how can we say that physics and mathematics are not dialectical? Some definition of dialectics, independent of Engels's own presentation of dialectics, is necessary in order to evaluate his claim, else we go around in circles. Let us therefore start with the assumption, which we consider to be valid, that Engels was basically right: the development of human society is dialectical, and dialectical materialism, utilizing the laws of dialectics, is indeed the science of society. What we need is the essence of dialectical development: it is irreversible, leading (by definition) to higher and higher forms, and it occurs through the development of internal contradictions. It is this understanding of dialectics which we shall use in evaluating Engels's claim that physics and mathematics exhibit dialectical behavior.

There is one other point necessary to make, before commencing the evaluation of Engels's work: I am a theoretical physicist, with the advanced academic training in physics and mathematics (for a Ph.D. degree) required for a successful research career. [I am now retired as a (Full) Professor of Medical Physics at Rush University in Chicago. My research, with some thirty peer-reviewed papers published in medical physics journals, has been concentrated on improving the radiation treatment of cancer using beams of electrons, photons (x-rays), and protons.] My training in physics and mathematics does not necessarily mean that my understanding of these fields is correct, but I am at least aware of their nuances, in a way that a person who only reads about them can never be. This advanced knowledge of physics and mathematics is the basis for my making such sweeping statements as that physics does not describe dialectical processes, or that mathematics contains no (logical) contradictions. Engels, for all his intellectual brilliance and contribution to human liberation, was so far as natural science goes always on the outside looking in, and this is why he was able to believe that dialectics, whose application to human society he so well understood, is also applicable to the physical sciences and mathematics.

B. Physics: Change of State of Matter

There are not very many ways in which one can construe the phenomena encompassed by the science of physics to behave dialectically. But there is one salient example, that of change of state of matter, which Guest presents in explaining Engels's first law of dialectics, the transformation of quantity into quality and vice versa:

> The simplest (and classical) example is the change of state [of] a substance, e.g. when a liquid becomes a gas (through boiling) or a solid (through freezing). Everyone knows that in such a case, gradual increase or decrease of temperature produces no departure from quality of liquidity until suddenly a point is reached where a complete transformation is effected. The liquid (as Hegel says) does not gradually become more and more gelatinous and semi-solid. It leaps at one bound from the liquid state to the solid.[24] [underlining added]

Now Guest is mainly quoting Engels's ideas here, and we shall have to see what Engels actually says on this question. But let us first evaluate this example from the standpoint of physics. What Guest is saying is simply wrong. Change of state of matter does not occur through change of temperature; change of temperature does not "produce" change of state at certain nodal points. Rather, the addition (or subtraction) of energy in the form of heat gradually changes the whole body from solid to liquid form (or vice versa); the body does not "leap at one bound from the liquid state to the solid". What is actually going on, for example, in the melting of ice is that water molecules which had previously been bound closely together into a solid substance gain the energy (heat) to break these bonds and dissociate from each other. As more and more heat is added, more and more of these bonds are broken and, as a whole, the substance becomes a liquid. Throughout this process the temperature remains constant.

One must understand that the temperature of an object is an *effect*, not a *cause*, of physical processes. Knowing the temperature of objects enables one to know certain ways in which they may interact with each other. For example, if two isolated objects of the same temperature are brought into contact with each other, nothing will happen, but if they have different temperatures then the colder one will absorb (heat) energy from the hotter one, at a rate proportion to their (instantaneous) difference in temperature, until they have the same temperature (intermediate between their two previous temperatures). The physical process occurring here is the exchange of energy in the form of heat, and the temperatures of the two objects is but a *measure* of what is going on. To say that raising the temperature of ice past the melting point "produces" a change to the liquid state, is like saying that changing the speedometer reading of an automobile from 50 mph to 60 mph causes the automobile to go faster.

Engels studied physics at great length, and he was quite well aware of the physical process involved in change of state. In AD, pp. 79-81, he goes into considerable detail about the transfer of heat in the change of state, and is quite clear on its being a gradual process (in contrast to Guest, who seems to be more imbued with Hegel). But Engels is also quite clear about regarding change of state to be an example of the law of transformation of quantity into quality and vice versa:

> This is precisely the Hegelian nodal line of measure relations, in which, at certain definite nodal points, the purely quantitative increase or decrease gives rise to a qualitative leap; for example, in the case of heated or cooled water, where boiling-point and freezing-point are the nodes at which – under normal pressure – the leap to a new state of aggregation takes place, and where consequently quantity is transformed into quality. [AD, p. 59]

And furthermore, in contradiction to his knowledge of the physical process involved, Engels states that "the merely quantitative change of temperature brings about a qualitative change in the condition of the water" (AD, p. 151, underlining added).

Thus, in this example, the problem is not simply Guest's interpretation of Engels, it is Engels himself. But let us return to

Guest, to the paragraph immediately preceding the one quoted above, to see what the stakes are:

> This law is essential for an understanding of the rise of *new* qualities, and also for understanding the quantitative effects which may follow the appearance of such new qualities. It is one of the fundamental superiorities of dialectical over mechanical materialism that the former understands how new qualities can arise at certain points of quantitative change – points where the change in quantity literally *becomes* qualitative change.[25]

Let us go beyond the content of the example which Guest then provides (about change of state of matter, which is simply wrong) to the form, and see whether this application might be useful if it happened to be physically valid. Is it true that "this law is essential for an understanding of the rise of new qualities", in this case change of state of matter? The answer here has to be no!, expressed far more emphatically than would be proper here. Not only is dialectics not essential here, it could not possibly provide better understanding of change of state of matter, for this process is *completely* understood by "mechanical materialism". And this example does not even attempt to provide *understanding* of the process of change of state: all dialectics does here is to make the observation that various forms of matter change state at particular temperatures, which is hardly a novel discovery. (This utter shallowness in the application of dialectics is reflected subsequently in the use of "Marxist-Leninist theory" by the international communist movement since the rise of Stalinism, as simply justification for whatever seems at the time to be appropriate to do.)

Change of state of matter is probably the clearest example which Engels provides of a dialectical process occurring in the field of physics, and let us examine it further to ascertain to what extent this process might be considered to be dialectical. Is this process, in terms of our own understanding of dialectical development, "irreversible, leading to higher and higher forms, and occurring through the development of internal contradictions"? Change of state of matter certainly doesn't proceed through the development of internal contradictions; there is no self-movement here, but rather the (completely determined) reaction to external forces, causing an increase or decrease of energy within the object.

And how about the irreversibility of the process, the gaining by the object of higher and higher forms, as implied by the law of the negation of the negation? While change of state is certainly a qualitative change, it is also reversible by an opposite physical process: we can melt ice by exposing it to warmer air, but we can then freeze the water right back up by exposing it to colder air. If we want to call melting the ice its "negation", then we would have to call the freezing process a "negation" (unless we were to give up all pretense of abstraction), so that the "negation of the negation" leaves us back where we started. Aside from being meaningless for practical purposes, such an interpretation of the law of the negation is hardly what we have in mind in our understanding of the development of human society – indeed, it is *because* we recognize that this development is dialectical that we would say that it is impossible for a feudal society, once having transformed itself to capitalism, to revert back to feudalism.

C. Physics: Other Examples

In attempting to apply dialectics to the physical sciences, Engels and his successors formulate (incipient) theories in terms of interpenetrating opposites. At best, such formulations are trivial, not furthering our understanding of the processes involved but at least not hurting our understanding (unless one decides to stop there, thinking that one actually knows something). For example, in Section A we have already seen the examples provided by Lenin[5]: action and reaction in mechanics (a branch of physics), positive and negative electricity in physics, the combination and dissociation of atoms in chemistry. The problem here isn't the recognition of dualities, such as two kinds of electric charge – it is their treatment as interpenetrating opposites, a mystical conception which leads us nowhere in understanding (nonliving) physical reality. In order to gain quantitative understanding of reality, (physical) scientists have had to develop their fields in terms of mechanical cause-and-effect relationships rather than dialectical mysticism, and the incredible success of "mechanical materialism" in developing the physical sciences is the ultimate proof of the inapplicability of dialectics to these fields.

But let us examine Engels's formulations when they are not trivial. In that case, because Engels doesn't really understand physics, and because he tries to use dialectics to overcome this lack of understanding (*i.e.*, to achieve qualitative understanding because he has not the option of achieving quantitative understanding), what Engels says is often muddled, and occasionally dead wrong. For example, Engels (DN, pp. 46-47) doesn't understand the difference between force and energy, and he tries to place them in terms of polar opposites, attraction and repulsion. In pp. 52-55, in trying to disprove a theory of the contemporary physicist Helmholtz, Engels decides that heat is a "repulsive force", acting in contradiction to the attraction of gravity and chemical forces. But physics is a quantitative science, and one cannot just play around with words (qualitative formulations) as Engels does; Engels is hopelessly lost in addressing Helmholtz's theory. Engels considers Helmholtz to be confused about the notion of force, and considers such confusion to be "the best proof that it [the notion of force] is in general not susceptible of scientific use in all branches of investigation which go beyond the calculations of mechanics" (p. 55), in particular physics and chemistry. Here Engels is not only centuries behind his time, but, because of his reliance on dialectics to understand that which can be understood only through mechanical materialism, he would throw out the whole science of physics. Since Newton's formulation of the laws of motion of matter, the notion of force has been integral to physics.

Engels makes other erroneous conclusions in DN. He doesn't understand the difference between momentum and energy, and wrestles with his confusion at considerable length (pp. 64-71). He states a "law of the indestructibility and uncreatibility of motion" (apparently corresponding to the law of conservation of energy) as expressing that "the sum of all attractions in the universe is equal to the sum of all repulsions" (p. 38), which is nonsensical, and goes on to formulate the (incorrect) thesis that bodies at very great distances from each other repel one another, instead of experiencing gravitational attraction (p. 259). He confuses entropy and energy, and in view of the (correct) law of conservation of energy, he dismisses (pp. 205, 216) the (correct) law of thermodynamics which states that the entropy of the universe is continually increasing.

Engels's understanding of motion of matter is outright mysticism, starting with his statement that "motion is the mode of existence of matter" (AD, p. 75). Such a conception has nothing to do with Newton's laws of motion, upon which the science of physics rests, nor could it be incorporated into a *quantitative* theory of physical processes. In Engels's view,

> Motion itself is a contradiction: even simple mechanical change of position can only come about through a body being at one and the same time both in one place and in another place, being in one and the same place and also not in it. And the continuous origination and simultaneous solution of this contradiction is precisely what motion is. [AD, pp. 144-145]

This is gibberish! Engels's view that a body can be in two places at once would obliterate the science of physics, just as his view that a body can at the same instant both be in a place and not be there is objectively an attack on human rationality. Engels actually has in mind the notion of qualitative change, and it is quite possible (and indeed necessary) to understand qualitative change of human society in the contradictory way which he poses, although certainly not as a logical contradiction. But we are here dealing with mechanical motion, which most assuredly is not a manifestation of qualitative change.

Coming back to our understanding of dialectical development, as being irreversible and leading to higher and higher forms, we see that there are examples from physics which might possibly be construed as exhibiting such motion. As examples of irreversible processes, we have that the flow of heat energy is always from a hotter to a colder body, and that the entropy of the universe as a whole is continually increasing. Entropy can be considered to be lack of order, so the universe as a whole is undergoing qualitative change, moving to "higher and higher levels" of disorder. A dialectical process defines the meaning for itself of "higher and higher levels", but increasing disorder would hardly seem to be what we mean by progressive dialectical development. Another example of irreversible qualitative development is the cycle of stars, from their birth to their death, but such development is completely preordained, like the motion of a clock once it is wound up; it is hardly useful to try to understand the cycle of stars in dialectical terms, for their development is well understood through the laws of physics (*i.e.*, through mechanical materialism).

And this is the problem with imposing dialectics on the science of physics: such imposition is completely mechanical and arbitrary, and does not aid our understanding of physical phenomena. In the field of chemistry, Engels gives (DN, p. 229), as an example of the law of transformation of quantity into quality, molecular oxygen (two atoms of atomic oxygen combined into a molecule) and ozone (three atoms of atomic oxygen). Well, it's true that molecular oxygen and ozone have markedly different chemical properties, but so what? How does thinking in terms of "the law of transformation of quantity into quality" help one to understand the difference in their chemical properties? Once again Engels is on the outside looking in, and his application of dialectics to the physical sciences only serves to trivialize dialectics and hinder us from understanding real dialectical processes. Physics in fact does not exhibit dialectical processes in any meaningful sense, and no competent physicist would attempt to impose dialectics on this science.

D. Mathematics

Let us be clear at the outset: mathematics is not a science. A science has as its subject matter a body of (natural) phenomena which can be understood in terms of an appropriate set of general laws, *i.e.*, a body of "qualitatively similar" phenomena. On the other hand, mathematics deals with logical systems each developed from a set of (unquestionable) axioms. It *may* happen that a particular set of axioms is such as to lead to a logical system which can successfully be used as an "analytic superstructure" to a particular science, for example the real number system with the usual arithmetic operations, or the system of calculus used in physics. Indeed, mathematicians have spent the great bulk of their time developing mathematical systems applicable to the sciences, for this has made their work practical to society (and therefore supportable by society); furthermore, natural phenomena have provided mathematicians with well-defined, meaningful problems to tackle. But in spite of the very close relationship between the sciences and mathematics, the two are of completely different character: a science is necessarily *materialist*, requiring continual experimentation and practical observation to verify its body of theory (which in turn is a summation of observed phenomena); but a mathematical system is *idealist*, with the validity of its propositions determined entirely by whether they flow logically from the (given) axioms of the system.

Thus, a mathematical system contains no (logical) contradictions – if the mathematical system you've developed from a particular set of axioms does contain contradictions, you've got to start over and do it right this time! (And if your mathematical system isn't applicable to a particular science, then you have to start again from scratch, with a new set of axioms.) With this understanding of mathematics, we see immediately that dialectics has no application to mathematics. Dialectical development occurs through the development of internal contra-dictions, but there is no self-movement in mathematics, and indeed there are no contradictions, internal or not.

Engels attempted to foist dialectical concepts onto mathematics, but it is clear from the passages on mathematics in AD and DN that he had essentially no grasp of the subject. His comments on algebra are painfully trivial and naïve, such as pp. 198-200 and 251-255 of DN, as well as:

But even lower mathematics teems with contradictions. It is for example a contradiction that a root of A should be a power of A, and yet $A^{1/2}=\sqrt{A}$. It is a contradiction that a negative quantity should be the square of anything, for every negative quantity multiplied by itself gives a positive square. The square root of minus one is therefore not only a contradiction, but even an absurd contradiction, a real absurdity. And yet $\sqrt{-1}$ is in many cases a necessary result of correct mathematical operations. Furthermore, where would mathematics – lower or higher – be, if it were prohibited from operating with $\sqrt{-1}$? [AD, p. 146]

To illustrate the law of the negation of the negation, Engels gives the following example:

> Let us take any algebraic quantity whatever: for example, a. If this is negated, we get -a (minus a). If we negate that negation, by multiplying -a by -a, we get $+a^2$, i.e. the original positive quantity, but at a higher degree, raised to its second power. In this case also it makes no difference that we can obtain the same a^2 by multiplying the positive a by itself, thus likewise getting a^2. For the negated negation is so securely entrenched in a^2 that the latter always has two square roots, namely a and -a. And the fact that it is impossible to get rid of the negated negation, the negative root of the square, acquires very obvious significance as soon as we come to quadratic equations. [AD, p. 164]

But Engels is just playing with words here, making no contribution toward understanding how to use algebra or toward understanding the nature of algebra, and in fact mystifying algebraic processes.

Regarding calculus, Engels fares even worse. Part of the problem seems to be his reliance on a book then eighty years old (AD, p. 470, footnotes 197 and 198), written some fifty years before calculus was placed on a rigorous footing with the introduction of the concept of limit by Cauchy. This study of Bossut's outdated book appears to have left Engels with his erroneous ideas on calculus, such as that curves and straight lines can be equated to one another (AD, p. 144; DN, pp. 200-201), and that 0/0 has mathematical meaning (AD, pp. 164-165 and 412-413). In fact, Engels points out that 0/0 represents a contradiction, but his response, based on his philosophical outlook, is acceptance:

> Therefore, dy/dx, the ratio between the differentials of x and y, is equal to 0/0, but 0/0 taken as the expression of y/x. I only mention in passing that this ratio between two quantities which have disappeared, caught at the moment of their disappearance, is a contradiction; however, it cannot disturb us any more than it has disturbed the whole of mathematics for almost two hundred years. [AD, pp. 164-165]

In contrast, it did disturb the great mathematician Cauchy, and he rooted out such contradictions from calculus, thereby making it possible to develop it into an advanced instrument for understanding nature.

Mathematicians have of course ignored Engels's efforts to foist dialectics onto their field, but we are fortunate in being able to examine a contemporary attempt to do so, albeit only at the level of algebra. In May of 1978, the Communist Labor Party held a philosophy seminar, and one of the conference papers which it subsequently published was devoted to dialectics in mathematics. We conclude this section by quoting a paragraph explaining "dialectical movement in equation-solving", to illustrate how mechanical and vacuous such attempts can be:

> The equation 3X+4=19 poses the contradiction "What is X?". In elementary school, children memorize the mathematical law: "When equals are added to equals the results are equal", an expression of dialectical interconnectedness of any equation. The movement which resolves the given contradiction begins by applying

this law and the equation -4=-4 to obtain 3X+4+(-4)= 19+(-4), or 3X=15. The oppositeness of addition and subtraction has been utilized to negate the 4, allowing it to move from one pole of the equation and penetrate the other, thereby changing both poles. A mutual penetration. We now use a version of the above law with "added to" replaced with "multiplied by" and the equation 1/3=1/3 to obtain 1/3x3X=1/3x15 or X=5. This time the oppositeness of multiplication and division is used to negate the 3 and again mutual penetration takes place resulting in the solution of the equation.[26]

E. Conclusion

It is most interesting to read what the Soviets had to say about dialectics two decades after the Bolshevik Revolution, and we can refer to the **Textbook of Marxist Philosophy**, prepared by the Leningrad Institute of Philosophy under the direction of M. Shirokov[27], in 1937. This work is a lengthy, very comprehensive review of Marxist and pre-Marxist philosophy, with particular application to the political struggles and questions confronting the Soviet state at the time. As was usual during the Stalin era and since, it reduces Marxist theory to justification, after the fact, of the political decisions taken, and one finds in it, for example, "proof" that Stalin's political opponents were wrong philosoph- ically, and that classes and class struggle are immediately to disappear under socialism. But it was also written after an unsuccessful attempt had been made to reformulate the natural sciences in terms of dialectics, as mentioned earlier[7], and what is most revealing is what the work does not say. The application of dialectics to mathematics is simply ignored, and, while lip service is given to Engels's views that dialectics is applicable to physics and chemistry, Shirokov has stepped very gingerly to avoid repeating Engels's obvious errors. Indeed, there appear to be only two clear-cut false statements in this area: that "a moving point" is simultaneously found and not found in a given spot" (p. 152), and that the development of physics and chemistry demanded a new methodological system, beyond mechanical materialism (p. 234). On the one hand it is good that Shirokov largely avoided Engels's erroneous views on the application of dialectics to the physical sciences and mathematics, but on the other hand it is most unfortunate that he could not openly identify Engels's errors, for by that time in the Soviet Union Marxist-Leninist theory had been transformed into a state religion in the service of a new exploitative class.

We trust we have demonstrated that the physical sciences and mathematics do not exhibit dialectical development, in the sense which we understand it: irreversible development, leading to higher and higher forms, and occurring through the development of internal contradictions. Thus we must forego Engels's view, so comforting and yet so wrong, that the dialectical development of human society is but one manifestation of the universal laws guiding all natural processes. Comforting indeed this view is, for we are assured thereby that the monstrous system of capitalism, like every other natural phenomenon, will necessarily come to pass, preordained to be succeeded by the higher form of communism – we are not idealists looking for Utopia, but have grasped the essence of all natural development!

So we cannot invoke universality of dialectical development. So what? The fact remains that human society, as well as all life processes, exhibits dialectical development in the sense understood by Marx and Engels, by Lenin and Mao. We do not need external justification for our observation that society develops dialectically; what we need is detailed understanding of this dialectical development. And such understanding, the raising of dialectical materialism ("the science of society") from an empirical science to a theoretical science, requires the systematization of dialectics, which is impossible to carry out so long as we remain muddled in thinking that dialectics is applicable to nonliving phenomena. Thus we must disabuse ourselves, once and for all, of the notion that all nature is dialectical.

REFERENCES

1. Karl Marx, *Theses on Feuerbach* (1945). Reprinted in an appendix to Ernest Fischer, **How to Read Karl Marx** (Monthly Review Press, New York, 1996), p. 171.

2. Mao Zedong, "On Contradiction", August 1937. In **Four Essays on Philosophy** (Foreign Languages Press, Beijing, 1968), p. 26.

3. *Ibid.*, pp. 27-28.

4. *Ibid.*, pp. 58-59.

5. Frederick Engels, **Dialectics of Nature** (International Publishers, New York, 1940), p. 26.

6. Mao, *Ibid.*, p. 65.

7. *Ibid.*, p. 67.

8. *Ibid.*, p. 55.

9. Mao Zedong, "Talk on Questions of Philosophy" (1964), in **Mao Tse-Tung Unrehearsed**, Stuart Schram, ed. (Pelican, London, 1974), p. 226.

10. Louis Althusser, "Contradiction and Overdetermination" (1962) and "On the Materialist Dialectic" (1963). Reprinted in Louis Althusser, **For Marx** (Verso, London, 1979).

11. Michael Albert, Leslie Cagan, Noam Chomsky, Robin Hahnel, Mel King, Lydia Sargent, and Holly Sklar, **Liberating Theory** (South End Press, Boston, 1986).

12. Margaret Randall, **Gathering Rage: The Failure of 20th Century Revolutions to Develop a Feminist Agenda** (Monthly Review Press, New York, 1992), p. 160.

13. Gayle Rubin, "The Traffic in Women: Notes on the 'Political Economy' of Sex". In **Toward an Anthropology of Women**, edited by Rayna R. Reiter (Monthly Review Press, New York, 1975), pp. 157-210.

14. Rosalind Petchesky, "Dissolving the Hyphen: A Report on Marxist-Feminist Groups 1-5". In **Capitalist Patriarchy and the Case for Socialist Feminism**, edited by Zillah R. Eisenstein (Monthly Review Press, New York, 1979), pp. 373-389.

15. Mao Zedong, "On Contradiction", August 1937. In **Four Essays on Philosophy** (Foreign Languages Press, Beijing, 1968), p. 28.

16. Frederick Engels, **Anti-Dühring** (Progress Publishers, Moscow, 1969), p. 15.

17. *Ibid.*, p. 16.

18. Frederick Engels, **Dialectics of Nature** (International Publishers, New York, 1940).

19. Joseph Stalin, **Dialectical and Historical Materialism** (International Publishers, New York, 1940), pp. 9-11.

20. V.I. Lenin, "On the Question of Dialectics". In **Collected Works**, Vol. 38 (Moscow, 1958), p. 357.

21. Mao Tsetung, "On Contradiction" (August 1937). In **Four Essays on Philosophy** (Foreign Languages Press, Beijing, 1968), p. 32.

22. David Joravsky, **Soviet Marxism and Natural Science, 1917-1932** (Columbia University Press, New York, 1961).

23. Vsevolod Holubnychy, "Mao Tse-tung's Materialistic Dialectics", *China Quarterly* **19**, pp. 3-37 (July-September 1964).

24. David Guest, **Lectures on Marxist Philosophy** (New Book Centre, Calcutta, 1963), p. 39. (First published in 1939 under the title **A Textbook of Dialectical Materialism**.)

25. *Ibid.*, p. 39.

26. Mike B., "Dialectics in Mathematics", *Proletariat* **4**, No. 4, pp. 33-37 (Winter 1978).

27. M. Shirokov, **Textbook of Marxist Philosophy**, 1937. Prepared by the Leningrad Institute of Philosophy under the direct of M. Shirokov, and reprinted by Proletarian Publishers, P.O. Box 40273, San Francisco 94140.

28. Albert Einstein, "Why Socialism?", *Monthly Review*, May 1949. Reprinted in Leo Huberman and Paul M. Sweezy, **Introduction to Socialism** (Monthly Review Press, New York, 1968), pp. 11-19

9/11 and Nationalism
Dave Jette, August 2018

This is a review of *9/11 Ten Years Later: When State Crimes Against Democracy Succeed* by David Ray Griffin (Olive Branch Press, 2011). The book demonstrates that the 9/11 attacks on the Twin Towers of the World Trade Center and on the Pentagon were an inside job, a false flag operation. The attacks did not, as claimed by the Bush administration and the mass media, result from a conspiracy carried out by operatives of al-Qaeda under the direction of Osama bin Laden.

An inside job

The idea that these horrendous attacks were carried out by elements of our own government is surely frightening. What is even more frightening is the total, mindless rejection by the great majority of Americans of this as an inside job that our enlightened country could never do. This appeal to nationalism as the basic faith of Americans is what I consider to be the most important understanding provided by the book, but let me first outline the author's proofs that there was a government conspiracy.

A coherent case relying on evidence

The bulk of this book is devoted to demonstrating, in great detail, the impossibility of the government's al-Qaeda conspiracy theory. Ten years after 9/11, Griffin has sorted through the relevant facts and opinions to produce a coherent case against the official explanation of what happened. The reader herself/himself will need to read the book to evaluate the veracity of the points he makes (pp. 27-50 for the WTC buildings) as well as the whole of his analysis. We can here mention only some of his salient points:

1. The twin towers were not brought down by the airliners which penetrated them. The official story is that the fires fueled by the jet fuel aboard the aircraft melted the steel supporting the top floors of each tower (and throughout the structure), causing the top floors to fall and take the rest of the building with them. But the melting point of steel is 2800°F, whereas these fires could not have been hotter than 1800°F -- the maximum possible temperature for hydrocarbon-based building fires. Previously, there have been no incidents of such fires bringing down steel-framed high rises, even in the case of examples cited by Griffin of much larger, destructive fires.

2. There were also huge horizontal ejections of steel from the twin towers, which could only have been caused by explosives, not by the fires.

3. A third World Trade Center building (WTC7) collapsed that day. It was very large (47 stories), and it came down some six hours after the twin towers (WTC1 and WTC2). It came down in the short time required for absolute free fall, which -- according to the laws of physics (conservation of momentum and conservation of energy) -- could not have resulted from the top floors collapsing and bringing down the remaining floors. (I happen to be a physicist, with a Ph.D. in theoretical physics, and I was able to calculate and confirm that the collapse time was too short for the building to have collapsed solely under the weight of the top floors.) Moreover, the building collapsed

directly into its footprint, symmetrically (straight down, with an almost perfectly horizontal roofline), which indicates that its base supporting steel columns were *all* destroyed at once.

4. Civil engineering experts thus concluded that all three WTC buildings were brought down by controlled demolitions. There was clear evidence of the presence of nanothermite (which can be tailored to behave as an incendiary or an explosive) in the WTC dust. In fact the WTC debris continued burning for six months -- impossible for ordinary building materials, but explained by the presence of nanothermite.

5. A number of calls reported from passengers on the doomed aircrafts were demonstrated to have been fake. Some calls couldn't have been made using cell phones at such a high altitude, and the seatback telephones had been made inoperative for a number of months by the airline.

6. The official story claims that a Boeing 757 flown by al-Qaeda's Hani Hanjour crashed into the Pentagon. However, his flight school judged Hanjour to be incompetent to fly even a single-engine aircraft. Whatever hit the Pentagon took an extremely difficult trajectory which would have tested even a highly experienced pilot. (But not to worry, for evidently the target was achieved: where the Pentagon was struck there were no high-ranking Pentagon officials. Instead, the personnel carrying out an investigation into several billion dollars of unaccounted funds, as well as the relevant records, were wiped out.)

7. Although Osama bin Laden immediately denied involvement in the 9/11 attacks and continued to do so, the US military "found" a video of him admitting responsibility. How they found it was never made clear, and it appears to have been bogus. Rather than capturing bin Laden and bringing him to trial, the U.S. military murdered him nine years later. By disposing of his body by burial at sea, they avoided any Muslim friends or acquaintances being able to identify him. (It's even probable that they killed someone else, for bin Laden had been in poor health and quite likely had already died.)

Griffin reports that while there is no consensus among the critics of the government's 9/11 conspiracy theory as to what actually hit the Pentagon, there is agreement that the WTC buildings were brought down by controlled demolition.

Obstructing an investigation, blaming al-Quaeda

The Bush administration did everything it could to prevent an investigation into 9/11. They appointed political operatives to executive positions in the 9/11 Commission (first attempting, unsuccessfully, to install Henry Kissinger as its Chairman!) and drastically underfunded the Commission's work. Vice-President Dick Cheney (not a Commission member) was, in fact, directly implicated in the attacks through his failure to order the military to shoot down the aircraft approaching the Pentagon. The 9/11 Commission produced the required report blaming al-Qaeda for the attacks.

The 9-11 Truth Movement

But as it became increasingly clear that the 9/11 attacks had been an inside job employing controlled demolition, the 9/11 Truth Movement was formed. It included many distinguished professionals from various fields (pp. 57-59 and pp. 222-226): former State Department officials, intelligence agency officials (including the CIA), and high-ranking military leaders as well as over 1500 architects and engineers in Architects and Engineers for 9/11 Truth. Their conclusion was that the attacks had been an inside job employing controlled demolition.

Failure of the mass media

However, the mass media in our country (but not elsewhere in the world) almost completely ignored the objections to the government's conspiracy theory. This is hardly surprising since our mass media as well as our federal government is controlled by the 1%, in whose interest was the subsequent invasion of and attacks upon oil-rich Muslim countries. (Some months before 9/11, our government had decided to create "regime change" in Afghanistan, Iran, Iraq, Libya, Somalia, and Yemen, and 9/11 gave it a widely accepted excuse to do so in its subsequent "War on Terrorism".) So mainstream opinion-makers commonly described members of the 9/11 Truth Movement as nuts, cranks, and idiots, as mindless advocates of conspiracy theories. (Of course, there are two conspiracy theories involved here, the government's blaming al-Qaeda and the 9/11 Truth Movement's blaming the government. But the mass media recognizes and dismisses derisively only the latter "conspiracy theory".)

The U.S. as the indispensable nation for good in the world

Also dismissing critics of the government's conspiracy theory out-of-hand were leftist intellectuals like Noam Chomsky, Alexander Cockburn, and Bill Moyers, as well as the great majority of Americans. Why would so many people refuse to examine seriously the evidence and arguments of the 9/11 Truth Movement? Chapter 8 of the book explains that what is operative here is nationalist faith. This (rather than Christianity, for example) is the basic faith of Americans: that the United States is the exception country, the indispensable country, for advancing civilization throughout the world. This faith rests on the belief that we are fundamentally a virtuous nation and, while we occasionally may make major mistakes (such as in Vietnam and Iraq), we are essentially good, never deliberately doing anything terribly evil. Surely our government could never kill some 2800 Americans for political purposes!

I admit that I succumbed to this outlook, at least partially. Right after 9/11, I was aware of criticisms of the official story, but I thought "who other than al-Qaeda might have carried out the attacks?".

Identifying a possible perpetrator

To answer this question, I listed five requirements for a possible alternate perpetrator: requisite technical expertise, political motivation (in terms of goals for the future), opportunity (ability to do what was practically necessary), expectation of being shielded from any exposure or retribution, and the will (ethics) to sacrifice so many people. With these criteria I deduced that the only possible alternate perpetrator would have to be ... Israel. I eliminated the United States since I didn't think that our government could possibly carry out such a heinous act against its own people.

But since then, I've read *JFK and the Unspeakable: Why He Died and Why It Matters* by James W. Douglass (2010). Douglas demonstrates definitively how President Kennedy was assassinated by what we term the Deep State (here, mainly the CIA but also including elements of the military and the Secret Service). Through Griffin's book I have now studied and understood the 9/11 attacks. I have concluded that it was not Israel (because of the complexity of the operation) but the "Deep State" that carried out the 9/11 attacks.

The Deep State and our nationalist faith

So where does this understanding leave us? We do have to be cognizant of the existence of the Deep State, rather than allowing it to be dismissed as just another "conspiracy theory" (derisively, in scare quotation marks). The Deep State intervenes on matters of critical importance to "national security," *viz.* the maintenance and expansion of the Empire.

More importantly, we have to understand the immense hold that nationalist faith has on Americans, in facilitating their support of our enormous military spending and of our imperialist wars throughout the world. We are well aware of the role of white nationalism (in the form of white supremacy) in our country, but the problem goes much deeper. An overwhelming nationalism pervades our society, evinced by the strong support for Donald Trump's call to "make America first" even among those who are not rabid white supremacists.

Some progressives will say that we shouldn't get bogged down in questioning the government's 9/11 story, that we should instead concentrate on more immediate matters facing us. But it is our duty to make people aware of what exactly is going on so that they can themselves decide what course of action to take, rather than deciding for them what information would be useful to them. If and when fascism comes to the United States, it will be on the basis of nationalism rather than simply white supremacy and conservative Christianity, and progressives should know what to expect and guard against.

Published in the August 2018 issue of "Works in Progress", Olympia, Washington

Relation of Progressives to the Democratic Party
Dave Jette, 11/27/2018

A. The Current Situation

1. Donald Trump is not simply the worst President in recent history – he is systematically bringing about fascism. For a detailed exposition on this, please see "Neo-Fascism in the White House" in the April 2017 issue of *Monthly Review* and **Trump in the White House: Tragedy and Farce** (Monthly Review Press, 2017), both by John Bellamy Foster. This fascism will be based upon White Supremacy, as seen most recently in Trump's reaction to the Charlottesville White Nationalist event, his support for retaining Confederate statues ("thing of beauty"), and his pardoning of former Sheriff Joe Arpaio.

2. In his inaugural address, Trump exclaimed in words which Hitler might have used:

From this moment on, it's going to be America First. ... At the bedrock of our politics will be a total allegiance to the United States of America. ... When America is united, America is totally unstoppable. ... Most importantly, we are protected by God. ... ...Together, We Will Make America Strong Again. We Will Make America Wealthy Again. We Will Make America Proud Again. We Will Make America Safe Again. And, Yes, Together, We Will Make America Great Again. (Quoted on page 9 of the Foster article.)

3. One reaction to the Trump agenda within the Democratic Party has been that it needs to swing (further) right, in order to win back Trump voters. However, there is also enthusiasm among active supporters of the Bernie Sanders campaign for trying to take over

the Democratic Party in order to have it advance a liberal agenda along the lines of the Sanders campaign.

B. Relating to the Masses

1. A great many people support decent living conditions for everyone. Progressives realize that this can be accomplished only through fundamental change away from our current capitalist system. Liberals believe it is possible to effect needed reforms without drastically altering our present economic system. Even self-identified conservatives will often support salient reform policies, especially if they are of libertarian bent.

2. In order to be effective, progressives have got to link up politically with liberals and many conservatives; without doing this, we have no chance of overthrowing the rule of the 1%. Such linkage will usually be of the form of mutual participation in particular, well-defined struggles for progressive social change, rather than broad agreement on necessary social change. Such mass struggle with progressives integrally involved is the necessary mechanism for transforming our society in a progressive direction. (Historically, the tendency not to associate with persons who don't fully accept the "correct line" has been a great weakness of the Left in our country, and it must be overcome.)

3. However, the electoral arena is also critically important, for, like it or not, that is where most people are at, in looking to bring about positive change in living conditions. This venue gives us the opportunity to propagate widely our ideas about necessary changes to make in our society.

4. Thus some progressives may choose to involve themselves in electoral politics, while realizing that fundamental transformation of our society can be achieved only through mass struggle. Electoral work can nonetheless play a critical supporting role by actually electing to office people who can implement progressive demands to some extent at least, and by tying together (ideologically) the various progressive struggles, so that people see the need to address all these issues as a whole, for their own liberation.

C. Relation to the Democratic Party

1. The Democratic Party has always functioned as the servant of the capitalist class, particularly (now) of the 1%. For untold decades its role has been to absorb and emasculate progressive struggles, and attempts to "reform" it or to take it over have always ended in failure. Recognizing this fact, many progressives may decide to have nothing to do with the Democratic Party, except as necessary to stave off the imminent prospect of outright fascism.

2. Other progressives may choose to run in elections as Democrats for tactical and/or strategic reasons. In primary elections, their opponents (mainstream Democrats} should be fought unrelentingly by progressives, as Bernie Sanders and Alexandria Ocasio-Cortez have demonstrated to substantial extent. The challenger should not hold back in offering a progressive alternative to what the mainstream Democrat stands for, even if so exposing the mainstream Democrat may weaken her/his chances in the general election. (A dubious, self-serving claim to make!) The progressive may even pull off an upset victory over the centrist, as has happily started to happen, but the point is that we

must not water down our progressive politics when we have a mass forum for expressing them.

3. Furthermore, after the progressive candidate (presumably) loses in the primary election, it no longer is the time to continue the attack on the mainstream Democrat's politics, for what then becomes foremost is the need to fight against the right-wing onslaught. So the progressive candidate may endorse her/his opponent's campaign and even use her/his campaign organization to get out the vote for that person, but only in the context (expressed publicly) that it is important to vote for the Democrat in order to defeat the Republican, not because the Democrat's politics are now beyond reproach.

D. Dealing with Progressive Democrats

1. There are definite limitations which genuine progressives running for or holding office as Democrats will encounter, in order to be able to work effectively with their Democratic colleagues, for example Sander's necessary endorsement of Clinton in spite of her dirty tricks. While we must push back against egregious politics, we also should cut progressive Democratic office-holders some slack, and concentrate on our important mass work. Realizing these structural limitations on progressive Democratic office-holders, we must not rely on them to effect the fundamental change needed in our society, but should welcome whatever support they are able to give to this effort.

2. In any case, the crucial point regarding relating to the Democratic Party is to avoid submerging progressive struggles to the needs of building that organization, as has occurred so often in the past to the acute detriment of the progressive movement. Doing this leads nowhere.

A New Approach to Supporting Progressive Electoral Candidates

Dave Jette, 6/6/19

Overview

Only a few years ago, the debate in Congress was over how drastically to eliminate Obamacare. Now the debate has turned around 180°, about replacing Obamacare with universal health care. Why is this? It is because, spurred by Bernie Sanders's forceful advocacy of "Medicare For All", new congresspersons and Presidential candidates have taken up this call; and because the American people have overwhelmingly realized that decent health care for everyone (especially themselves) is what they really want. That is, through forceful advocacy we have been able to break through the propaganda of the health insurance industry and "free" many politicians from their economic domination.

The situation is similar for a number of salient issues: liberal and "progressive" politicians will give lip service to important causes, while restraining themselves from necessary action because of the influence of immense campaign contributions. What we propose to do is to promote electoral campaigns which go all the way in advocating what is necessary, what should be wholeheartedly supported by Americans once the blinders of the politicians and the mass media are removed. For example, the "Green New Deal" is now unexpectedly gaining major traction.

At the same time, we need to avoid the trap of trying to link all these struggles together to build a vision of our progressive society of the future. This is certainly relevant to do in another context, but it would be a diversion for this new approach. Rather, we should concentrate separately on each of these most critical points, not providing the opportunity to stray into larger questions which can draw away one's attention.

Salient Issues

1. Climate Change. We all know how terrible the threat to our environment is. We must reject half-hearted efforts to combat this threat, regardless of the costs. The Green New Deal is a fine starting point for this.

2. Nuclear War. This threat to extinguish human society is little publicized, but it's a toss-up of which is going to destroy us first, this or global warming. All the nuclear powers are glad to increase the effectiveness of their nuclear arsenal, and Trump has been accelerating this doomsday process with huge research funding and by abrogating agreements with Russia to mitigate the likelihood of nuclear war. It is absolutely imperative that we bring this threat onto the table, and demand reversal of the buildup of the ability to wage nuclear warfare.

3. Economic warfare. Using its vast economic power, our country continues to try to dominate the world economically, particularly through economic sanctions which are illegal under international law. We have got to stop these efforts at "regime change" against countries which we can't control, such as Iran, North Korea, Venezuela, and Cuba. Rather, we must respect their sovereignty and maintain normal trade relations with them. Overall, we must recognize the need for an economically multipolar world rather than continue to pursue a unipolar one controlled by our country.

4. Military warfare. We must stop waging war throughout the world, and in particular we must immediately withdraw our armed forces from the Middle East and Afghanistan and stop giving military support to efforts to crush civilian uprisings against tyrannical regimes, as in Yemen. This will entail drastic cuts in our military spending, whose proceeds can then be used for needed domestic programs. And we don't need to use words like "imperialism" and "aggression", which just serve to add fuel to the flames – all that is needed is to combat our imperialism, concretely.

5. Universal health care. Proper health care is a human right, and we must do whatever is economically necessary to ensure that all our people receive such. As is used in every other industrialized country in the world, this will require a government-run program for funding health care, replacing the huge waste (and huge profits) inherent in our private health insurance industry. Various "Medicare For All" proposals provide a solid start for this transition.

6. Poverty. We must take seriously the vast amount of poverty in our country, which is all the more outrageous given our great wealth. We must take responsibility for building our country through slave labor followed by many decades of Jim Crow, as well as through the genocide and land-grabbing of its indigenous people. Concrete programs with major economic cost will be necessary to turn the country around, and reparations of some sort may be required.

7. Palestine. Israel is an apartheid state, with different laws for different people, and especially since 1967 it has been steadily absorbing the remainder of Palestine at the expense of the indigenous people, through illegal colonization. The United States bears great responsibility for Israel's outrageous behavior toward the Palestinian people, for it is alone in automatically supporting whatever Israel does. Because of the power of the Israel Lobby in Congress, as well as the reluctance of many Jewish liberals to threaten the existence of Israel as an apartheid state, Israel has been able to get away with doing whatever is feasible, and we have the particular responsibility of demanding unequivocally the cessation of the oppression of the Palestinian people. Specifically, we must demand:

    (a.) No economic, military, or diplomatic support of
        Israel so long as it oppresses the Palestinians.

(b.) Unqualified support for the international
BDS ("Boycott, Divestment, and
Sanctions") Movement, which has been
called by Palestinian civil society as a
nonviolent means of forcing Israel to
comply with international law.

(c.) Support of the "right of return" for
Palestinians who were ejected from their
homes as the Zionists colonized Palestine.

What *Not* to Include

1. The Progressive Litany. We do have to make clear that,
generally, we are a part of the overall progressive movement:
we shall not support any electoral campaign which does not
support basic human rights, such as reproductive rights and
LGBTQ rights and opposition to the vilification of
immigrants and Moslems But it won't be useful to call for
adherence to the whole progressive litany – this would just
be a red herring.

2. Capitalism vs. socialism. Socialism is such a vague term
that is bandied about that discussion of the need to replace
(present) capitalism with some sort of socialism just detracts
from our message of concentrating on the (seven) salient
issues. Best not even to mention these terms.

3. White supremacy. Nope, the problem of white supremacy
in our society has been brought to the forefront, and it won't

do any good to give people an excuse for defensiveness. Combatting white supremacy is implicit in addressing the issue of poverty – no need to even mention these words.

4. *The 1% and corporate control of our political system.* We all know that this is the root problem, but it won't further our agenda to bring it up.

5. *Christian conservatism.* Many conservative Christians are genuinely concerned with the welfare of people and not just getting into Heaven themselves. We must try to engage them in these concrete struggles – especially concerning poverty – rather than automatically placing them on the other side of the line. We must seek to engage seriously with them over saving our country.

6. *Trump and fascism.* Of great importance, of course, but bringing it up would just serve as a red herring for us.

What We Should Do

For this project, we need to create a Political Action Committee (a "PAC") which identifies and financially supports electoral campaigns which promote our demands for addressing our salient issues. Of course, there will be different emphasis placed on these issues; while all of them are critical for federal election campaigns, some won't be relevant for local ones.

This PAC should be run by a Board of Trustees
consisting of a group of veteran activists who have well
demonstrated their political judgment and organizational
competence. They will make the final decisions about
funding, of course, as well as how the organization functions
– a good number of Trustees would perhaps be eight. The
PAC should build up a network of volunteers around the
country who can identify worthwhile electoral campaigns
and candidates and make sure that they are in harmony with
our goals. It may also be necessary to have some salaried
personnel. We should be thinking in terms of dealing with
millions of dollars, and have systematic fundraising (and
endorsements).

I have been advocating for supporting progressive
electoral candidates for many years, mainly in supporting
alternatives to the Democratic Party, which like the
Republican Party is closely controlled by the 1%. The
formulation which I have been using is that there are two
basic reasons for running progressives for office:
    (a.) to elect progressives to office to actually
            implement progressive policies; and
    (b. to generalize (ideologically) particular
            progressive struggles to the whole range of
            such struggles, so that people see the need for
            an overall transformation of society.
However, in light of the ideas presented in this "New
Approach" proposal, I now think that the second goal is
presently not of primary importance, and for it I would

substitute encouraging and supporting the self-organizing of people to address effectively the seven salient issues identified here.

The point is that we have got to get out of the trap of relying on politicians – especially mainstream Democratic ones – to solve the fundamental problems of our society. Rather, we have got to get used to taking these issues into our own hands, dragging along reluctant politicians as well as possible. It is only through doing this will it be possible to effect fundamental progressive change in our society. Thus this proposal is aimed toward supporting (financially) electoral campaigns which undertake mobilizing people around one or more of these salient issues. And while it won't be necessary for a campaign to emphasize all of these salient issues, the fact that they receive substantial support from our PAC will tie them at least somewhat to all of these issues.

Author's Recent Political Activity:

Electoral Campaigns

1996: Secretary, Washington State Campaign for Democracy (supporting Ralph Nader's 1996 Presidential campaign in Seattle)

1999: Treasurer, Curt Firestone for Seattle City Council

2000: Treasurer, Joe Szwaja for U.S. Congress

2001: Treasurer, Curt Firestone for Seattle City Council

2003: Secretary, Larry Gossett for King County Council (successful)

2003: Treasurer, Brita Butler-Wall for Seattle School Board (successful)

2003: Treasurer, Brent McMillan for Seattle Monorail Board

2006: Treasurer, Aaron Dixon for U.S. Senate

2007: Treasurer, Joe Szwaja for Seattle City Council

2008: Treasurer, Divest from War (Seattle initiative against Israeli occupation of Palestine; terminated early by court order)

2010: Treasurer, Richard Curtis for U.S. Senate (aborted early)

2012: Treasurer, Sue Gunn for U.S. Congress (in Olympia)

2014-16: Treasurer (2014-15) and Secretary (2016), Washington State Coalition to Amend the Constitution (successful statewide Initiative I-735 to overturn the Supreme Court's *Citizens United* decision and get big money out of politics)

2016: One of two consultants to Jill Stein's Green Party Presidential campaign, to ensure that FEC regulations were being followed

2018: Treasurer, Stonewall Bird for U.S. Congress (in Bellingham)

Electoral Organizations

1989-90: Corresponding Secretary, Washington State Rainbow Coalition (arose from Jesse Jackson's 1988 Presidential campaign)

1999: Secretary, Seattle Progressive Coalition

2001: Secretary, Green Party of Seattle

2006-08: Secretary, Green Party of Seattle

2007-08: Deputy Treasurer, Green Party of Washington State

2008: Treasurer, Green Senate Campaign Committee (of national Green Party)

2008: Treasurer, Progressive Action Committee (short-lived fundraising committee based in Seattle)

2010: Treasurer, Washington State Progressive Electoral Coalition

2011-17: Treasurer, Justice Party (arose from Rocky Anderson's 2012 Presidential campaign)

2012-13: Treasurer, Justice Party of Washington State

2014-17: Treasurer, Green Party of Washington State

2015: Treasurer, Green Party of Seattle

Other

2003: Treasurer, Peace Action of Washington

2011-present: Trustee, Tom Warner Workers' Defense Fund which donates approximately $70,000 annually to progressive organizations.

2018: Became Managing Trustee of Workers' Defense Fund.

Author's Scientific Career:

The author received a B.A. degree in mathematics at Cornell University in 1964, an M.S. degree in physics at Brandeis University in 1967, and a Ph.D. degree in theoretical physics at the University of Calgary in 1970. After postdoctoral training in medical physics at the University of Chicago in 1977-79, he joined the Department of Therapeutic Radiology at Rush University in Chicago in 1980, eventually rising to (Full) Professor of Medical Physics there in 1993. He also served as Research Director of the Northwest Medical Physics Center in Seattle in 1985-87, as Director of the Division of Therapeutic Radiological Physics at the Institute of Applied Physiology and Medicine in 1987-1993, as Executive Director of the Lawrence H. Lanzl Institute of Medical Physics in Seattle from 1993 to 2003, and as Vice-President and Research Director of ICT Radiology in Livingston, New Jersey from 2008 to 2010.

At Rush University the author taught graduate courses in medical physics and mentored successfully three Ph.D. students in medical physics. During his scientific career he published 32 peer-review articles, two of which won the annual Farrington Daniels Award for best paper on an aspect of radiation dosimetry published in *Medical Physics*. His research concentrated on improving the treatment of cancer through radiation beams. He attained membership in Phi Beta Kappa, Phi Beta Phi, and Phi Eta Sigma scholastic honorary societies and in Sigma Xi scientific research society, and obtained professional certification by the American Board of Radiology in Therapeutic Radiological Physics.